California's Botanical Landscapes

A Pictorial View of the State's Vegetation

Michael G. Barbour
Julie M. Evens
Todd Keeler-Wolf
John O. Sawyer

California Native Plant Society

Within the Russian Wilderness of the Salmon Mountains, the Duck Lake area — greatly extolled in lectures and texts by John Sawyer—contains the richest diversity of conifer species (17 conifers) in California. Jeff Bisbee

Previous pages:

1 *Tidy-tips* (Layia platyglossa) *occurs in expansive displays in the Arena Plains Unit of the Merced National Wildlife Refuge in the northern San Joaquin Valley.* Gary Zahm

2 *Conway Summit, within the BLM's Bridgeport Valley management area, has a diverse mosaic of wet meadows, big sagebrush* (Artemisia tridentata)*–bitter brush* (Purshia tridentata) *shrublands, and aspen* (Populus tremuloides) *groves, here with fall color.* Robert Wick

3 *California black oak* (Quercus kelloggii) *woodland near Cloverdale, Sonoma County.* Reny Parker

2707 K Street, Suite 1
Sacramento, CA 95816
(916) 447-2677
www.cnps.org

Library of Congress Control Number: 2015948046

ISBN 978-0-943460-55-0

Produced with care by Collaborative Publishing Services • Bozeman, Montana • Jane Freeburg, Editor/Designer
www.collaborativepublishingservices.com

Printed in the United States of America • First Printing

Michael E. Kauffmann

Dedicated to John O. Sawyer

John O. Sawyer, born and raised in California, pursued graduate work in the Midwest, the New World and Old World tropics. He returned to the state in 1966 to begin a long and distinguished teaching and research career as a professor of botany at Humboldt State University. He taught a number of subjects, including plant geography and plant taxonomy labs, but perhaps became best known for his field courses in plant ecology and vegetation. John guided hundreds of undergraduate and graduate students toward professional paths in botany, plant ecology, and vegetation science, including two of the authors of this book.

Additionally, John's role in compiling statewide information about vegetation is what we all thank him for and most remember him doing. In 1991 John stepped forward when asked by the CNPS Vegetation Committee to lay down the "new paradigm" of California vegetation classification. He became the lead author of the two editions of *A Manual of California Vegetation*, with a clear vision of how and why those books should be written. Later he interacted with the National Panel on Vegetation of the Ecological Society of America, helping to incorporate his broad expertise in the classification of North American and global vegetation.

John had a strong sense of beauty, whether appreciating a piece of classical music, a sweeping Turner landscape, or the perfect arrangement of glacially scoured canyons and massively deposited boulders emerging out of a subalpine meadow. John's appreciation and passion for natural landscapes and the vegetation they supported was infectious. All of us who knew him will remember his exclamations of joy as he left the bed of an old logging road or a crowded trailhead, climbed over a ridge, and viewed the wilderness beyond. His aesthetic appreciation of landscapes and vegetation and profound obsession with deciphering the universe led him to such a successful and influential career. As a student of nature he could find something interesting about any piece of land, whether desert, alpine, coastal, or montane. John was a gifted observer and a careful and meticulous scientist. As an academic, he combined his sense of aesthetics with his disciplined mind. Of the several books and many scientific articles he wrote in his career, many are deeply rooted in aesthetics and his love of nature's beauty. He knew that through definition and description would come the most defensible way to protect and revere what he loved about the natural world. He and his students not only quantified the beauty of vegetation by classifying it, but went even further to quantify more subtle aspects of vegetation, defining the characteristics of old growth forest or carefully detailing the natural history of his favorite landscape of all, the Klamath Mountains.

John was an ardent conservationist. He served as a CNPS President, founding member of the CNPS Vegetation Committee, and later a CNPS Fellow. He was on the board of the Save the Redwoods League, and he volunteered much of his time to other similar conservation duties. During the last year of his life he was a strong supporter of this book, working on several chapters prior to his death in August 2012. It seems entirely fitting, therefore, that we dedicate this book—an attempt to meld the beauty and the science of California's botanical landscapes—to the memory of John O. Sawyer, his spirit, and his indefatigable love of vegetation.

Todd Keeler-Wolf
Julie M. Evens
Michael Barbour

Understory of a giant sequoia (Sequoiadendron giganteum) *forest with Pacific dogwood* (Cornus nuttallii). Richard Turner

Contents

Opposite: Frying-pan poppy (Eschscholzia lobbii) *and sky lupine* (Lupinus nanus) *along a rocky volcanic plateau of the North Table Mountain Ecological Reserve.* Julie M. Evens

Foreword

Many people have heard of Shangri-La, the secret Himalayan valley of beauty and tranquility in English novelist James Hilton's 1933 fantasy, *Lost Horizon.* Fewer have heard that after Hilton visited the Trinity River Gorge in northwest California's Klamath-Siskiyou Mountains some years later, he said it seemed like a *real* Shangri-La. I didn't know about Hilton's comment in 1971 when I first hiked into what was then the Salmon-Trinity Alps Primitive Area just north of the gorge. But I would have agreed with him. I'd never experienced anything quite like the sense of secluded calm that its landscape evoked: granite peaks; glacial lakes; ancient groves of pines, firs, and incense cedars; meadows full of wildflowers with common names as colorful as their inflorescences—great red paintbrush, rein orchid, grand collomia, elephant's head, green corn lily. The Alps seemed to shelter an inexhaustible variety of beautiful plants—the closest *I'd* come to an earthly paradise.

A few centuries ago, California as a whole probably was as close to Shangri-La as any place on Earth. Isolated by ocean, desert, and mountain, it was a largely peaceful land of people who lived well enough on a year-round abundance of nutritious seeds, roots, bulbs, herbs, and fruits that they didn't need to cut down its forests or plow up its meadows.

That peaceful isolation is gone, and civilization has pushed much of California's native vegetation aside. Exotic weeds have reduced or replaced it in large areas: introduced diseases threaten many species. Yet vibrant native associations survive in much of the state, and the longer civilization is here, the more we realize their value. When I first hiked the Trinity Alps, for example, botanists John O. Sawyer, Dale Thornburgh, and Petra Bingham were discovering that the Russian Peak area a short distance to the north has among the world's highest levels of conifer diversity, with seventeen species of pine, fir, spruce, cedar, hemlock, yew, and juniper within a single square mile.

The New Hampshire-sized Klamath-Siskiyou region has become internationally known for its biodiversity, yet it is one of the *less* famous of California's floral regions. North Coast redwood forests and Sierra Nevada giant sequoia groves are the most spectacular of these, but many other California plant species and associations are globally celebrated for their outstanding qualities.

California's desert—North America's driest and hottest—has uniquely diverse and resilient floras in its Mojave, Sonoran, and Great Basin regions. The central and southern coast, with the continent's only "Mediterranean" climate, has a mosaic of woodland, grassland, and chaparral so intricate and rich that it contains some of the world's highest numbers of endemic species. Even the Great Valley, where civilization has pushed native flora hardest, retains precious remnants of the grassland, wetland, savanna, and riparian woodland complex that John Muir called "the grandest and most telling of California's landscapes...one smooth, flowery, lake-like bed of fertile soil...one furred, rich sheet of golden compositae..."

World famous as California's flora is, it needs to be much better known and appreciated—especially *in* California—if civilization's relentless growth is not to push it even further aside. People need to better understand that the patterns of native vegetation—and the geological and climatic conditions influencing them—are what *define* the land they inhabit. They need to see more than "woods," "bushes," and "grass." They need to see the beauty and diversity of the thousands of species that have evolved here over millions of years, and the dynamic complexity of the ways those species associate to form landscapes, as with the Trinity Alps' peaks, lakes, groves and meadows.

California's original people thrived here for many thousands of years by knowing its plants. Today's Californians should take a lesson from that. This lively, comprehensive, beautifully illustrated compendium of articles by botanists working to understand and protect native vegetation is a vital step in that hopeful direction.

David Rains Wallace

*Opposite: California pitcher plant (*Darlingtonia californica*) fen in the Deadfall Lakes area at the headwaters of the Trinity River.* Robert Wick

The Minarets from Lake Ediza with mountain hemlock (Tsuga mertensiana)*, whitebark pine* (Pinus albicaulis)*, sedges* (Carex *spp.*)*,*
Sierra gray willow (Salix orestera)*, and dwarf bilberry* (Vaccinium cespitosum)*.* Jeff Bisbee

In 1991, CNPS established a "Plant Communities Committee," an all-star team united in an important mission and driven by an ambitious vision. They believed that if we could develop standardized approaches for identifying, describing, and naming vegetation, then we would all have a powerful tool for understanding and conserving our most beloved plants and places. The rest, as they say, is history.

You hold in your hands an exciting part of that history. In just 24 years, CNPS volunteers, staff, and partners established the first standard approach for characterizing and mapping California's vegetation (*A Manual of California Vegetation*), and developed professional training workshops to teach the new approaches. As a wealth of new field data was collected, vegetation ecologists in CNPS and the California Department of Fish and Wildlife identified and defined finer levels of vegetation, and completed an award-winning second edition of the *Manual*. Today Californians have mapped in detail over one-third of the state and are working to map the rest. We are approaching the day when we will finally, thankfully, have a definitive treatment and mapping of all of California's hugely diverse and wonderfully special vegetation patterns!

As you enjoy this beautiful book, please consider the marvel it represents. In less than a quarter century, a small determined community developed the tools and then performed much of the work to document and map our nation's most diverse and heterogeneous ecosystems. They developed information that is key to understanding and defending these unique places—and they took some pretty great photos at the same time! This book is testimony to our ability to accomplish great and important feats. It's a continuing story of success that should inspire us all as we look to the future and consider how we each, and together as a Society, can make a real difference. It is an invitation to join the thousands of Californians who unite, as CNPS, in celebrating and saving the beautiful native plants of our state and the natural landscapes in which they grow.

Dan Gluesenkamp
CNPS Executive Director

Acknowledgements

A book of this type requires patience and cooperation among many. First and foremost, we thank the 14 contributing authors for their time and devotion to this project. The responsibility of putting forth the concept of the book has been on our set of shoulders. We have relied upon the accumulated knowledge and dedication of our co-authors for their written prose; without their perseverance, this book would not have been realized.

We also wish to thank the leadership of the California Native Plant Society for giving us enough time and funding to advance the knowledge and understanding of our state's invaluable botanical landscapes, and the many fine photographers who have graciously shared their high quality photos with us. First off, we thank Jeff Bisbee, Michael Kauffmann, and Robert Wick for their great generosity in donating so many marvelous photographs. Secondly, several other photographers have donated multiple photos that we express gratitude for: Jim André, Jennifer Buck-Diaz, Mike Dolan, Forest Gauna, Deborah Hillyard, Don Lepley, Connie Millar, Kyle Merriam, Craig Odegard, Andrea Pickart, Robert Preston, Gina Radieve, John O. Sawyer, Allan Schoenherr, Teresa Sholars, Rich Spjut, John Sterling, John Stuart, Pete Veilleux, Wallace Woolfenden, and Gary Zahm. More generally, we thank all of the photographers who have contributed to this project. The photographs in this book are of equal importance to the text that they accompany.

Another less visible but prominent group includes the support staff who largely worked to contact authors, photographers, help select photos, scan slides, and prepare the manuscripts for review by the editors and the designers. Chief among them is Jaime Ratchford. Jaime has worked tirelessly to ensure selection of photos, review plant names, review text and caption revisions, and cross-reference all parts of this project moving toward completion. We also had help from other CNPS program staff, Josie Crawford, Kendra Sikes, and Deborah Stout, and volunteers, including Olga Béguin, Catherine Curley, Maureen Dannen, Abigail Genelsa, Martin Rose, and Caroline Tippets. California Department of Fish & Wildlife (CDFW) state staff, Aicha Ougzin, and Esri staff, Charlie Frye and Charles Convis, helped us in producing the California map figures. CDFW regional staff, Jennifer Olson and Michael van Hattem, helped review animal names. Our copy-editor and designer, Jane Freeburg, has brought the raw material into a stunning and tastefully assembled opus. Our prepress specialist, Joanne Bolton, has brought extra light and liveliness to the photographs. Thanks to them, we have gotten the book produced!

We are thankful for the generosity of the following sponsors of the book: the North Coast Chapter of CNPS, Tejon Ranch Conservancy, The Nature Conservancy, Ron Maertz, and Steve Schoenig.

Additionally, this book was made possible because of the tremendous support from two anonymous donors, Esri, and California Wildlife Foundation/California Oaks, whose generosity is greatly appreciated.

We are grateful to our families for supporting us in burning the midnight oil and in accompanying us on field trips to myriad places to obtain specific photographs. Lastly, we want to acknowledge John O. Sawyer, our co-author, mentor, and friend who passed away while this book was in its early stages, yet not before he left his indelible mark of reverence to nature.

California Native Plant Society

Todd Keeler-Wolf
Julie M. Evens
Michael G. Barbour
September, 2015

Introduction

After the publication of the second edition of *A Manual of California Vegetation* (Sawyer et al. 2009) the three co-authors felt compelled to condense the information in that 1,300-page tome into a book that more simply extolled the splendor and virtues of California's vegetation. The "MCV2," as we call it, is replete with ecological information, botanical facts, and distribution details. However, it does not have a single photo of any of the nearly 500 vegetation types described. We wrote the MCV2 as a resource for those already initiated into the world of vegetation classification and California ecology and biogeography. We decided not to add even more pages with photos to that already expanded text, since we wanted the book to be somewhat portable and it would have been extremely difficult to select just a few representative photos. Following publication in late 2009, we asked ourselves: how can anyone, aside from a determined biologist, wade through that much information and come away delighted and captivated by the diversity of California vegetation?

We hope the book you now hold can incite an initial foray—and invigorate a deeper interest in—the beauty and complexity of our state's vegetated landscapes. This book came about because of our own passion for the subject. We feel that there is something beautiful and compelling about vegetation patterns which stand on their own aesthetically, regardless of the interests that scientists have about vegetation, or the needs that natural resource managers have in using vegetation classification as a tool of their trade.

In reflection, we see that vegetation appreciation needs an official public launching. During our more grandiose moments while writing this book, we have envisioned it as inspiring groups of natural history "vegetation buffs" in admiring these "living landscapes," and relating them back to the plants, animals, and the physical settings they also love. The small, portable Peterson Field Guides and UC Press California Natural History series have several books summarizing trees, shrubs, wildflowers, grasses, and plant life which have engaged many amateur and professional natural historians. However, the depth and complexity of California's vegetation deserves an even larger book to visually inspire and not merely summarize and simplify the subject.

One of the great rewards of delving deeply into the subject of vegetation for much of our careers is gaining the satisfaction and understanding of "how it all fits together." We know so much more now about the state's vegetation than we did 20 years ago—through its description and comparative analysis; detailed mapping of its patterns; and relating those to patterns of climate, topography, soils, and natural processes such as fire and drought. We are putting this deeper knowledge to use in many ways. Describing and mapping vegetation can accurately reflect human and natural changes. Vegetation patterns are, when well understood and defined, narrators of the living landscape: the record of current and past conditions, and invaluable markers in measuring and monitoring the change, quality, and biological potential of almost any piece of California.

We realize that we can't bring all of you instantly into this long-enduring passion of ours without some sort of come-on. How can anyone expect a stranger who has not previously experienced a combination of complex realizations to become instantly fascinated? Vegetation science is a synthetic field, which requires understanding of plants, their reproductive and physiological biology, and their genetic relationships. In addition it requires a comprehension of earth sciences (soils, weather, climatology, geology, geomorphology, hydrology, and fire) and biogeography. However, unlike some other synthetic sciences, vegetation has a great advantage—it is easy to see, smell, touch, and even hear (consider, for example, wind blowing through a pine forest or an aspen grove). Wherever we go, in so many ways and so many areas, it adds enormously to what we consider the inherent beauty of our California landscape.

Our model for this book is the successful and beautiful publication of *California's Wild Gardens*, first published by CNPS in 1997 and reprinted in 2005. That book has

Opposite: Rancheria Creek, a tributary of the Merced River, incises down a slope of oaks (Quercus *spp.*) *and pines* (Pinus *spp.*), *with a mixture of white alder* (Alnus rhombifolia), *red willow* (Salix laevigata), *and western redbud* (Cercis occidentalis). Julie M. Evens

The upper Sacramento River incising the oak laden foothills near Red Bluff. Robert Wick

a biogeographic approach capturing the state's botanical gems and "botanical hot spots" to emphasize the myriad special settings within the state boundaries. Likewise, we have subdivided California into ecologically distinct regions; e.g., the Sierra Nevada, the Klamath Mountains, the Great Valley and the warm Southern California Coast, and we bring out the diversity of vegetation that inherently varies between these climatically and physically different areas. "Ecoregions" are conceptual and synthetic in nature, much as is vegetation. They are defined by combinations of climate, topography, geology, and yes—vegetation (Bailey 1995). In fact, vegetation is typically the only biotic component used to define ecoregions because it can be mapped and comprehended at what many ecologists call a "landscape scale."

What is "landscape scale"? A landscape is commonly defined as "all the visible features of an area of countryside or land, often considered in terms of their aesthetic appeal." Relatively speaking, a landscape is bigger than a small single watershed like a creek flowing through a suburban park. A landscape can contain several to many creeks, some small, some larger. The canyons or valleys these watercourses flow through have internal variations, some steep and some gradual, some underlain by acidic granitic rock, some by basic limestone, or harsh serpentine. These watersheds head in mountains of different heights and exposures—steep, shady, rocky, exposed. Within these landscapes, maps of vegetation can be created and viewed much as maps of geology, soils, topography, or precipitation. And within these landscapes and using such maps, we can see the patterns and relationships between the physical and biological features.

The Chapter Authors

We are fortunate that knowledgeable and influential vegetation ecologists abound in California. Their origins and backgrounds are different, but interconnected. Beginning in the 1960s biological science experienced a spike in academic interest around ecological research with the rapid growth in the science of ecology and of the understanding of ecosystem dynamics and processes. Several universities began hiring young plant ecologists and natural community ecologists, among them Michael Barbour and John Sawyer. The influence these men and their colleagues had upon students and published literature soon found its way into agencies and organizations that manage and protect the natural processes. These included federal and state agencies, as well as NGOs with multi-species conservation interests.

The US Forest Service was the first agency in California to adopt an ecosystem-based management system. It required definition and description of natural vegetation

Private landowners, TNC, and CDFW have integrally conserved lands in the Lassen Foothills. Julie M. Evens

Patterning on south- versus north-facing slopes in the Central Coast Ranges include grasslands versus oak woodland. Todd Keeler-Wolf

and the understanding of natural potential vegetation and variation of timberlands and other renewable resources across California's 20 million acres of Forest Service lands. Regardless of its basis, the publication of high quality quantitative descriptions and definitions characterizing vegetation on forest lands was a strong foundation for further vegetation classification. Three of our authors are products either directly or indirectly of these ecosystem management years.

Picking up on the Forest Service lead, other state and federal agencies along with non- government land management and conservation organizations also hired plant ecologists. The California Department of Fish and Wildlife (CDFW; formerly Fish and Game) developed a new classification of state scientists: Plant Ecologists. Two of the authors came from this background. Similarly, the California Native Plant Society and The Nature Conservancy (TNC) also hired such people to undertake their ecosystem-based conservation and land management philosophies. Local educators from colleges have also been avid contributors to the field of plant and vegetation ecology by bridging relationships between these and related fields of conservation biology, and natural history, often going well beyond their own communities to emphasize the importance of these topics. Every author on this team has experience in writing both scientific and popular accounts of nature and vegetation. It is important to us for them to be able to describe their topics from an enthusiast's viewpoint.

Appreciating Patterns

One of the first things we notice about vegetation is the characteristic repeating patterns it forms across a landscape. We look at a view from a ridge in the inner North Coast Ranges and see consistent patterns of scrubby oaks and pines intermingling on rocky and shaded north-facing slopes, while on the opposite, sunny south-facing slopes are dense, low aggregations of shrubs such as chamise and ceanothus. On a closer scale at the edges of the Great Valley, we can see distinct differences between stands of purple needle grass and California poppies on low rounded hillocks alternating with thick turfs of creeping rye grass and native clovers in concave adjacent draws. These patterns exist because of the shift in environments between the settings. Speculating about the underlying reasons for these patterns gets us to think about the fundamental requirements of different plant species and improves our understanding and our ability to "read nature."

As we travel from place to place around the state we may see the same plant species occur in different settings. For example, in the western Mojave

Creosote bush (Larrea tridentata), *brittle bush* (Encelia farinosa), *and white bursage* (Ambrosia dumosa) *intermix along the slopes, alluvial fans, and inselberg of the Silurian Hills.* Julie M. Evens

Desert we might see stands of white bursage on rugged badlands of eroded alluvial fans facing south, while in the Riverside Mountains of the Colorado Desert, we might see white bursage only on sheltered north-facing slopes. These differences tell us about the breadth of tolerance of the component plants in vegetation. When the desert is relatively wet and cool, white bursage prefers well-drained sunny slopes, but when it's relatively hot and dry, white bursage prefers shadier cooler exposures. We can muse for hours about the environmental controls of vegetation and how widely predictable a given type of vegetation is before its local landscape or soil preferences change. These types of observations attune us with the patterns and processes in nature and thus improve our appreciation of its complexity.

How Chapters Are Written

The chapters in *California's Botanical Landscapes* are organized by different ecological regions (or ecoregions) of the state because we think that you as a reader will most likely relate to a sense of place. We emphasize a discussion of the landscape more than the individual plants by accentuating the vegetation patterns, rather than diversity at the species level, and we also underscore the relationships between these patterns and the other physical patterns. Even though it is possible to see the same or very similar types of vegetation in several of the major ecoregions, we feel it's important to note which areas have unique and distinctive vegetation. For example, the Mojave and Sonoran deserts both have creosote bush scrub, but you can only see Joshua tree woodland in the Mojave and microphyll woodland in the Sonoran.

We have allowed each of the authors the license to describe their area as they saw fit. But all were encouraged to use some of the terminology of vegetation classification such as "alliance" and "stands," as well as structural terminology such as "emergent layer" and "cover."

The Value of Terminology, Whether Large, or Small Scale, Singly or Collectively

Although this book is about "vegetation," it is impossible and moreover unwise to consider botany, the study of plants and plant species, in any simple way separate from vegetation ecology. We can study the ecology of an individual plant—how it reproduces, its pollination strategy, its nutrition, climatic requirements and physiological limits. In vegetation ecology we are extending those concepts beyond the individual species to groups of species that tend to coexist in a given location. A widely used synonym for vegetation or "vegetation type" is the term "plant community" —a predictable assemblage of certain plant species growing under similar environmental conditions .

The "plant association" is considered the finest scale or "fundamental unit". One can envision small numbers of species of plants growing close together to form a local "association" of plants. Currently in California we recognize almost 2,000 plant associations (about ⅓ the number of full species of plants known from the state). In this book we typically use the level just above the plant association in the chapter accounts. This is the "alliance." Although the term evokes some sort of arranged "agreement" amongst plants that suggests a conspiracy or accord, it is a term that was coined by Europeans to imply a somewhat broader concept than the plant association, as in "an alliance" of nations.

If we are on the ground and walk into a small verdant mountain meadow, we can see discrete patches containing a mixture of different species of sedges, rushes, grasses, and maybe leafy herbs such as shooting stars, lilies, monkey-flowers, and wild onions. Each patch of plants equates to a separate stand of a given association, and the meadow is made up of many separate associations, which may belong to the same or different alliances. We can see these patterns of the different stands and relate them to differences in soil moisture, or soil texture, slope position, site disturbance, and shading. Imagine if we are not on the ground, but could only view this meadow from an aerial photo taken several thousand feet above the ground. Much of that meadow would simply look green in the wetter parts and paler in color or even tawny in the drier parts. We may know there are several types of alliances in that meadow and because of the similarity in the moisture gradient, they sort into related groups of overlapping species: dry, moist, and wet meadows. This is something we can map. Because these general conditions have a physiological as well as a phenotypic similarity, it is also valuable information to map. Thus, even if we could not see each individual alliance in the drier part of the meadow, or know for sure what plant associations were representing each of those alliances, we could discern the "group" level.

The variable "zoom telescope" view of vegetation (from individual plant, to small groups, up to much broader groups of ecologically related vegetation) is not only a valuable and natural way of interpreting nature, but keeps the mind nimble and endlessly questioning. It deepens our appreciation and expands our philosophical, emotional, and biological connection with nature.

How to Establish a Life-long Appreciation of Vegetation

Do you need to know the names of all the alliances and associations, and their relationships to one another, to recognize and value vegetation? The answer is no. Vegetation patterns are what we all inherently know about most of our favorite landscapes. Could it be that low coastal ridge covered with California oat grass, Douglas iris, and blue-eyed grass overlooking the Pacific, or that riotously colorful display of wild lupines, phacelias, gilias, and poppies on the sunny slopes in the foothills overlooking the Valley? Is it the delicate interconnected swales colored brilliantly gold from banks of goldfields and tidy-tips in the Valley vernal pool landscape? Or the surprisingly thick and shrubby understory of huckleberry and wild rhododendron under a North Coast mixed evergreen forest? Is it that sweeping arc of blue-green oak woodlands sprinkled across the low rolling foothills of the Sierra Nevada and Central Coast Ranges, or that intricately woven carpet of chamise and manzanita chaparral along steeper slopes of these same regions? Is it that weirdly unfamiliar and singular Joshua tree woodland amongst the pale granite monoliths of a moonlit night in the Mojave Desert highlands? Or the surprise when we round a bend in the stark desert canyon to see a tightly packed grove of desert fan palms? Is it the insectivorous tiny twinkling sundews or the cobra-like pitcher plants within an emerald montane fen?

Redwood (Sequoia sempervirens) *forest in Prairie Creek Redwoods State Park.*
Gina Radieve

Meadows, coastal scrub, and oak-madrone woodlands in view along Prewitt Ridge, Big Sur coast. Pete Veilleux

There are innumerable such scenes awaiting the curious observer in California. Each holds the potential to both please and fascinate.

Photography of Vegetation and Travel

One of the most obvious ways to gain a deeper appreciation of vegetation is through photography. Unlike wild animal photography or close up botanical photography, you do not need special lenses, tripods, strobes or a large degree of patience to photograph most vegetation. Vegetation is "our size," it is usually most easily viewed from a mid-range—not too close, not too far. Although it is more difficult to capture the variety of small herbaceous communities or capture the depth and perspective of the interior of an old growth Port Orford cedar forest, it is fascinating to simply photograph vegetated landscapes and see the patterns "emerge" as you look at them later.

The more we travel, the more we can wonder. One of the greatest challenges about vegetation is developing a relative and regional consciousness about similarities and differences between types of vegetation. As we stand looking over a deep canyon in the coastal mountains of the Big Sur coast, it may be difficult to believe that the narrow ribbon of redwood at the bottom of a steep ravine is representative of completely different global vegetation than the broad-leaf woodland of oaks and madrones immediately adjacent to it. However, if we look widely for where redwood grows most extensively throughout its range, it shows up in areas with cool temperate climates compared to where most of the oaks and madrones grow, in areas with warmer and usually somewhat drier climates categorized as warm temperate/Mediterranean. The fine-grained topographic partitioning of vegetation in coastal California relates to the microclimates associated with this landscape. Indeed, on a very local scale the redwoods in that ravine do have a much cooler and moister climate than the sunnier, warmer, and less well watered oak-madrone woodland. These amazing juxtapositions are commonplace in our physically diverse state.

Once we know of the origins of vegetation we can reflect more deeply on the possible mixings of plants and plant communities under varying scenarios of climate change—what might happen or what is unlikely to ever happen in the composition or the physical arrangements of the plant communities of the future. Regardless, we need to ensure that our landscapes stay connected, so that the movement of plants and related biota can continue to shift in space and time.

Michael G. Barbour
Julie M. Evens
Todd Keeler-Wolf

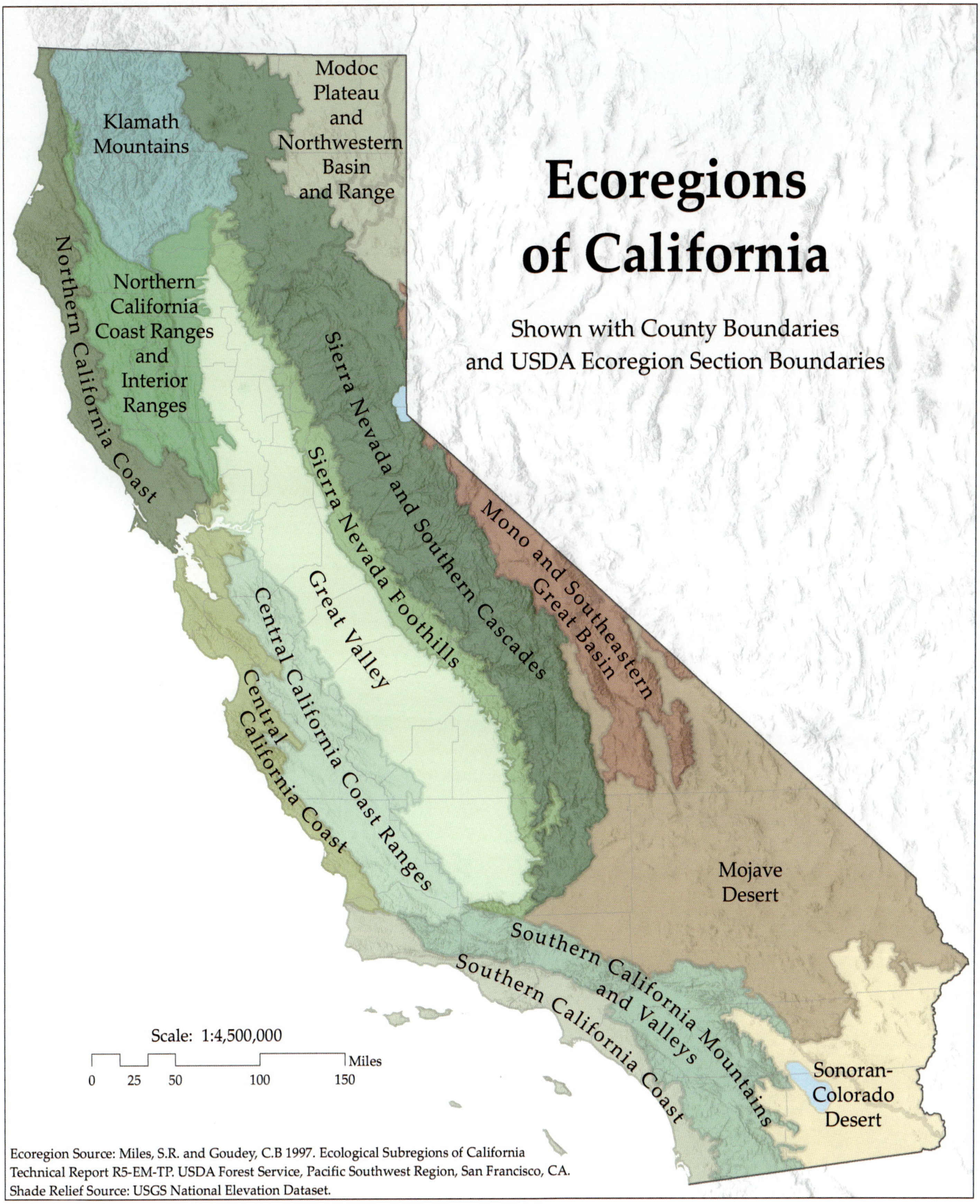
Ecoregions of California
Shown with County Boundaries and USDA Ecoregion Section Boundaries
Modoc Plateau and Northwestern Basin and Range
Klamath Mountains
Northern California Coast
Northern California Coast Ranges and Interior Ranges
Sierra Nevada and Southern Cascades
Sierra Nevada Foothills
Great Valley
Mono and Southeastern Great Basin
Central California Coast Ranges
Central California Coast
Mojave Desert
Southern California Mountains and Valleys
Southern California Coast
Sonoran-Colorado Desert
Scale: 1:4,500,000
Miles
0 25 50 100 150
Ecoregion Source: Miles, S.R. and Goudey, C.B 1997. Ecological Subregions of California Technical Report R5-EM-TP. USDA Forest Service, Pacific Southwest Region, San Francisco, CA.
Shade Relief Source: USGS National Elevation Dataset.
Charlie Frye & Charles Convis/Esri

CHAPTER ONE

Northern California Coast

California's North Coast is renowned for its world-class scenery, attracting many thousands of visitors annually. Rainforest levels of precipitation fall over long winters, alternating with summer drought ameliorated by frequent fog. Remoteness and a reputation for gray skies (and large earthquakes) have kept population levels far lower than along other coastlines. Thus, when the sun does shine, it reveals a dramatic coastline relatively unscathed by the kind of shoreline build-out typical elsewhere. For the purposes of this chapter, the northern California coastal strip extends from the Oregon border south to Marin County—a narrow area only a few kilometers wide that covers the immediate coast. More inland parts of the North Coast region are discussed in Chapter Two, Northern California Coast Ranges and Interior Ranges.

The North Coast's restless geology produces a geomorphically diverse landscape that in turn supports many different vegetation types. From Cape Mendocino north, the Cascadia subduction zone erupts in mega-earthquakes at 300–500 year intervals, causing instantaneous subsidence of estuaries, localized up-thrusts of coastal terraces, and tsunamis that engulf beach and dune systems. South of the infamous Triple Plate Junction off of Cape Mendocino, the San Andreas Fault's mostly horizontal movement results in scarps,

Top: Douglas fir-covered peaks rise above summer fog at remote Shelter Cove on the Humboldt County coastline.
Robert Wick

Above: The steep northern California coastline limits wide beaches to relatively few areas. Here, seastacks rise through the beach sand near Trinidad, Humboldt County.

Opposite: The Lost Coast is aptly named for its remoteness. Highway 1 was routed inland along this steep stretch of wilderness.
Andrea Pickart photos

Top: Coastal dune willows take on fall hues amidst darker beach pines in the deflation plain, the zone where the freshwater table is close to the surface and can form seasonal ponds later in the winter.
Andrea Pickart

Above: Distinct vegetation types occur in zones parallel to the shoreline at the Lanphere Dunes. From west to east: the foredune; narrow semi-stable parabolic dunes; deflation plain; and the unvegetated transgressive dunefield that is slowly covering the stabilized, forested dunes.
Dave Kenworthy

ridges, and troughs. The fault line is expressed as the trough that forms Tomales Bay, separating the Point Reyes Peninsula from the mainland.

The landscape of the North Coast has also been shaped by the climate changes induced by glacial cycles. The botanically rich stair-step marine terraces of Mendocino are the result of Pleistocene sea level stands combined with seismic uplift. This medley of geologic influences allows for extraordinary diversity in the vegetation of the North Coast.

Coastal Dunes

The North Coast is home to some of the most diverse and intact dune systems remaining in North America. Extensive Quaternary dune systems are found at Crescent City (Tolowa Dunes), Humboldt Bay/Eel River, the Ten Mile River (MacKerricher State Park), Point Arena (Manchester State Park), Tomales Bay, Point Reyes Peninsula, and Bodega Bay. The vegetation of the most northern dune systems is more aligned with dunes of Oregon and Washington than with more southerly California dunes.

In common with dune systems of the Pacific Northwest, the Humboldt Bay and Tolowa Dunes represent two different mobilizations of sand. Older, stabilized, forested dune features to the east are being slowly engulfed by a more recent pulse of sand, possibly generated by the last great Cascadia earthquake of 1700. Ten Mile Dunes once had a more extensive dune forest, but it had already been reduced to fragments by the time noted botanist and geologist William S. Cooper painstakingly explored and documented the dunes of the California, Oregon, and Washington coastlines in the 1920s and 1930s.

Dune Forest

Northern California's dune forests are thought to be relicts of a once-continuous band of largely closed-cone conifer forests that stretched along the California coast before the Hypsithermal—a relatively arid period that ended 5,000 years ago. The present-day forests occupy isolated, moist, cool pockets separated by subsequent contraction during climate cooling. These forests now function as refugia for species adapted to cooler temperatures. Several species found in the forests of the Lanphere Dunes on Humboldt Bay represent disjunct populations of species more typical of higher elevations, including the recently discovered little prince's pine (*Chimaphila menziesii*), a diminutive plant normally found at this latitude only above 700 m.

The most common type of vegetation on the older, stabilized dunes of Humboldt and Del Norte counties is the beach pine (*Pinus contorta* ssp. *contorta*) forest alliance. These forests, found just a few kilometers west of the towering and better-known redwoods (*Sequoia sempervirens*), have a moody and fantastical air. Epiphytic lichens such as net lichen (*Ramalina menziesii*) drape the limbs of the stunted pines that must eke a living

from a thin mantle of soil overlying the porous and nutrient-deficient dune sand. Where the canopy is open, the forest floor is often carpeted by the dwarf shrub kinnikinnick (*Arctostaphylos uva-ursi*), along with luxuriant mosses such as broom moss (*Dicranum scoparium*). This green carpet is punctuated by snowy tufts of the coastal reindeer lichen (*Cladina portentosa* ssp. *pacifica*). Tucked away in secret corners of the forest are strange symbionts such as candy stick (*Allotropa virgata*) and spotted coralroot

Top: Beach pine forest on stabilized dunes.

Above and inset: Coastal reindeer lichen forms snowy tufts alongside kinnikinnick in open areas of the beach pine forest alliance at the Lanphere Dunes.

Andrea Pickart photos

Top: *Sunset at Ma-le'l Dunes Cooperative Management Area with Sitka spruce in the path of a migrating dune.*

Center: A gray fox in the forest.

Above: Spotted coralroot.
ANDREA PICKART PHOTOS

(*Corallorhiza maculata*). Elegant native grasses are also found here, including fragrant vanilla grass (*Hierochloe occidentalis*) and tall trisetum (*Trisetum canescens*). Where the canopy is closed, a dense understory of shrubs can make the forest virtually impenetrable. Evergreen huckleberry (*Vaccinium ovatum*), salal (*Gaultheria shallon*), and coastal silk tassel (*Garrya elliptica*) are all common, thicket-forming shrubs. The berries provided by these shrubs serve as an important food source for mammals including the gray fox (*Urocyon cinereoargenteus*), as well as for neotropical migrants whose movements are timed to coincide with berry production.

The beach pine forest is also home to other coniferous trees, and when they become locally common they are recognized as associations within the alliance. Sitka spruce (*Picea sitchensis*) is able to germinate and grow in the shade of beach pine, and attains much greater stature. The forest relies on the windthrow caused by strong winds to generate gaps needed for beach pine regeneration. Grand fir (*Abies grandis*) is also found locally, often in the sheltered "ravines" that were left behind between adjacent advancing parabolic dunes. Douglas fir (*Pseudotsuga menziesii*) and a few western hemlocks (*Tsuga heterophylla*) are also found in the forest, along with the broad-leafed trees madrone (*Arbutus menziesii*) and cascara buckthorn (*Frangula purshiana* ssp. *purshiana*).

Low-lying areas in the dune forest support swamps that differ subtly from those found outside the dunes. These swamps are created where the winter water table comes close to or rises above the ground surface. Swamps that are only saturated or ephemerally flooded can still support beach pine, which has a tolerance for moist soils. But here the forest understory changes from nearby upland pine forest. Slough sedge (*Carex obnupta*) forms a continuous groundcover in places. Occasionally, the leather grape-fern (*Botrychium multifidum*) sends up its bizarre yet

beautiful fertile fronds. Lower, wetter swamps support coastal dune willow (*Salix hookeriana*) thickets and wax myrtle (*Morella californica*) scrub.

Within the dunes are a few perennial streams, fed by springs and runoff from the porous dunefields. Along these watercourses the vegetation shifts from conifers to red alder (*Alnus rubra*) forest, providing for a sharp ecotone that is attractive to wildlife of all kinds. These riparian swamps have dense understory thickets of salmonberry (*Rubus spectabilis*) and California blackberry (*Rubus ursinus*). Other understory species include pink-flowering currant (*Ribes sanguineum* var. *glutinosum*), coast twinberry (*Lonicera involucrata* var. *ledebourii*), and wood ferns (*Dryopteris arguta* and *D. expansa*). A touch of color is provided by the purple flowers of coast hedge nettle (*Stachys chamissonis*).

The small dune forest remnants at Ten Mile Dunes have a different composition from the stabilized dunes to the north. The closed-cone bishop pine (*Pinus muricata*) is found with grand fir and tanoak (*Notholithocarpus densiflorus*). These remnants hint at a once more extensive bishop pine dune forest when the stabilized dunes were more distant from the shoreline; conifers found in the more exposed dune swales today are limited to beach pine.

Dune Scrub

South of Point Arena (itself a transitional area), older, more stable dunes support dune scrub rather than forest vegetation. Silver dune lupine–mock heather (*Lupinus chamissonis–Ericameria ericoides*) scrub can be found on relatively stable dunes at Point Reyes, Tomales Bay, and as small relicts at Bodega Bay Dunes. Associated shrubs include coyote brush (*Baccharis pilularis*) and yellow bush lupine (*Lupinus arboreus*). The latter has been introduced to coastal dunes north of Sonoma County. This community sports bright colors in summer, with

Top: Coastal dune willow thickets form at the margin of the migrating dune-field where rainwater, percolating through the open dunes, drains toward the forest.

Above: Intertwining, moss-covered limbs of wax myrtle above a floor of slough sedge in a low-lying area of dune forest (wax myrtle scrub). Andrea Pickart photos

Top: Blooming silver dune lupine north of Abbott's Lagoon, Point Reyes. The dense vegetation in the foreground and middle ground comprises silver dune lupine–mock heather scrub.

Above: Advancing dunes engulf the beach pine forest, leaving the bleached skeletons of Sitka spruce and beach pines in their wake. Andrea Pickart photos

blue- and yellow-flowered lupines joined by the subshrub seaside woolly sunflower (*Eriophyllum staechadifolium*) and the deep red stalks of flowering dudleya (*Dudleya farinosa*). The rare dune gilia (*Gilia capitata* ssp. *chamissonis*) is found in this community at Point Reyes and Tomales Bay Dunes.

Migrating and Semi-stable Dunes

The true drama of the dunes takes place at the boundary between the older (forested) and newer (migrating) dunefields. Here, the steep slipfaces of advancing dunes engulf forests and swamps. As the vegetation dies, the thin mantle of soil is dislodged and the older dunes become cannibalized by the new advancing front, reworking the old sediments into new geomorphic features. There is little vegetation on these actively migrating dunes; plants are unable to take hold in such unstable substrate.

However, closer to the ocean newer dune features have formed from the interaction of wind, newly arriving sand, and vegetation. Sea lyme grass (*Leymus mollis*) patches occur on the foredune, a shore-parallel ridge that forms just above the beach. However, foredunes are not always present on Northern California coasts: some dunes, such as those found

at Gold Bluffs Beach, are limited to a low hummocky topography above the backshore. On high energy shorelines, the foredune is periodically breached by wind, forming blowouts. When blowouts migrate inland they take the form of narrow parabolic dunes, which leave behind vegetated "trailing ridges" that flank the deflating floor of the dune as it migrates. The result is a ridge-and-trough topography that can be seen in its natural state at the Lanphere and Ma-le'l Dunes.

The herbaceous alliance found on these more recent upland dune ridges is called dune mat (*Abronia latifolia–Ambrosia chamissonis*). Vegetation on these newer features is often patchy, with areas of open sand. Soils are poorly developed or absent, so these dunes are referred to as "semi-stable." The annuals, perennials, and subshrubs that make up the dune mat community are characterized by numerous adaptations for life in shifting sands, summer drought, infertility, salt spray, and wind-stressed conditions. Dune mat is a very diverse vegetation type. The disturbance and stresses of the dunes assure that succession is held in check, although some areas inevitably give way to forest (or, from Sonoma County south, to dune scrub).

Annuals such as smooth-seeded cryptantha (*Cryptantha leiocarpa*), the rare dark-eyed gilia (*Gilia millefoliata*),

Top: Yellow sand-verbena blooms in the low hummocky dunes above the backshore at Gold Bluffs Beach.

Above: Windblown sand accumulates on a foredune colonized by sea lyme grass patches.

Andrea Pickart photos

Top: A naturally stabilizing parabolic dune (right) supports dune mat on its trailing ridges. Adjacent to and underlying the advancing dune is the vegetated trough of an older, stable parabolic dune (left) supporting herbaceous wetland vegetation.

*Above: The beach sagewort association of dune mat at MacKerricher Dunes, includes beach sagewort, beach bur-sage (*Ambrosia chamissonis*) and seashore bluegrass.*
ANDREA PICKART PHOTOS

and the endangered beach layia (*Layia carnosa*) emerge in winter months when many of the perennials die back above ground. The annuals are able to complete their life history before most perennial species begin flowering. The endangered Menzies' wallflower (*Erysimum menziesii*) is one of the first perennials to bloom, signaling the arrival of spring with its charismatic yellow flowers in late March and April. As wallflowers fade, beach pea (*Lathyrus littoralis*), completely invisible in its dormant state below ground, emerges and fills large areas of open sand with its violet and white blossoms. As spring turns to summer, waves of color succeed each other as yellow sand-verbena (*Abronia latifolia*) and purple-flowered seaside daisy (*Erigeron glaucus*) give way to the pink pompoms of beach buckwheat (*Eriogonum latifolium*), and by late summer, the golden tints of dune goldenrod (*Solidago spathulata*). Interspersed with the colorful wildflowers are perennial rhizomatous and stoloniferous grasses, especially seashore bluegrass (*Poa macrantha*), and on slightly more stable ridges, red fescue (*Festuca rubra*).

A distinctive association of dune mat, characterized by beach sagewort (*Artemisia pycnocephala*), occurs throughout the North Coast dunes, and is actually the dominant association at all but Humboldt Bay Dunes. This community is distinguished by lower cover—as well as lower species diversity—than other associations. Although lacking some of the showier perennials found elsewhere, this subtype is quite beautiful during

peak flowering of beach sagewort, with its finely dissected, silvery foliage and pale golden inflorescences. Commonly co-occurring are beach knotweed (*Polygonum paronychia*), California sea thrift (*Armeria maritima* ssp. *californica*), seashore lupine (*Lupinus littoralis*), beach buckwheat, and seashore bluegrass. Annuals include footsteps of spring (*Sanicula arctopoides*), cream cups (*Platystemon californicus*), and, at Ten Mile Dunes, the rare roundhead Chinese houses (*Collinsia corymbosa*). Other North Coast rare or endangered dune endemics include sand dune phacelia (*Phacelia argentea*) found in the Tolowa Dunes of Del Norte County, Howell's spineflower (*Chorizanthe howellii*) at Ten Mile Dunes, curlyleaf monardella (*Monardella undulata*), and Tidestrom's lupine (*Lupinus tidestromii*) at Point Reyes.

Dune Deflation Plains and Basins

In the low troughs between vegetated trailing ridges (called deflation basins), as well as in the more expansive deflation plains that form behind migrating dunefields, vegetation shifts to seasonal wetlands known as swales or slacks. These wetlands support the slough sedge swards alliance, in which a number of freshwater to slightly brackish wetland species co-occur, such as Pacific silverweed (*Argentina egedii* ssp. *egedii*), fringed willowherb (*Epilobium ciliatum* ssp. *watsonii*), springbank clover (*Trifolium wormskioldii*), yellow-eyed and blue-eyed grass (*Sisyrinchium californicum*, *S. bellum*), and the showy orchids coast piperia (*Piperia elegans* ssp. *elegans*) and hooded ladies' tresses (*Spiranthes romanzoffiana*). Brewer's rush (*Juncus breweri*) forms monotypic stands in areas that are transitional between wetland and upland, but floristically these stands are classified as dune mat. The swales rapidly succeed to coastal dune willow thickets and—from Ten Mile Dunes north—eventually to beach pine forest.

Invasive Dune Species

Semi-stable and migrating dunes have been the target of reclamation efforts

Top: Slough sedge swards intermix with the winter branches of coastal dune willow in the flooded deflation plain at the Ma-le'l Dunes.

Below: A cluster of hooded ladies' tresses grow among slough sedge in a swale at Lanphere Dunes.

Andrea Pickart photos

Above: Dense coastal dune willow thickets near the mouth of the Mad River provide resting habitat for migrating songbirds.

Below: European beach grass swards at the Ma-le'l Dunes.
Andrea Pickart photos

in California since the late 1800s. The City of San Francisco is built over a massive, once mobile dunefield. Stabilization efforts began with European beach grass (*Ammophila arenaria*), brought to San Francisco in 1868 and soon after planted in other northern California dunes. European beach grass, which has denser shoots and can trap more sand than the native sea lyme grass, reduces aeolian transport of sediments inland. The foredune becomes steeper and overstabilized as the processes supporting native dune plant communities are curtailed. In Humboldt County, seeds of yellow bush lupine were collected from the Presidio and first planted in 1908 along the railroad corridor on the North Spit. Yellow bush lupine is native from Sonoma County south, where it is a constituent of native dune scrub.

However, in far northern California and the Pacific Northwest, there are no native dune scrub communities. When fast-growing yellow bush lupine shrubs become established in native, herbaceous dune mat, they quickly outcompete native species. At the same time, their nodulated roots and fast turnover rate cause the soil to become nitrogen-enriched, allowing secondary invaders such as ripgut brome (*Bromus diandrus*) to establish and persist even after the lupine is removed. In some areas where yellow bush lupine was introduced, dune vegetation has succeeded to novel assemblages of both native and non-native species, including dense stands of coyote brush growing with nested polypody (*Polypodium calirhiza*). Hottentot fig (*Carpobrotus edulis*) has a less studied introduction and invasion history, but is now present throughout northern California. Hottentot fig has hybridized extensively with the non-invasive sea fig (*C. chilensis*), forming a hybrid swarm throughout its range.

The spread of invasive dune plants has led to the recognition of several semi-natural alliances, including European beach grass swards, ice plant (*C. edulis* or other ice plants) mats, and yellow bush lupine scrub. Although invasive species were introduced to every major northern California dune system, aggressive protection and restoration have resulted in large expanses of native vegetation at most of these sites. Twenty years after restoration at the Lanphere Dunes, species diversity and cover is comparable to intact dunes that were never invaded.

Coastal Terraces and Sea Stacks

Along the coast of Mendocino County lie a series of sea-cut terraces that give rise to a sequence of very interesting and often unique vegetation types. The bedrock is almost entirely greywacke sandstone. Localized variation is created by distance from the ocean's windblown salts and fog. So while the climate and the bedrock are essentially the same, the vegetation is dramatically different depending on the age of the terraces. This can be seen most clearly from the Ten Mile River to the Navarro River, where a fault block has uplifted terraces. In some places these terraces have remained flat, which in turn has allowed the soil to age. The soils on the surrounding slopes are continually renewed by natural erosion from upslope, resulting in higher fertility. If the slopes are far enough away from the ocean and salt spray, they are vegetated by the dominant regional redwood forest vegetation.

This terrace sequence occurs in very few locations on the Mendocino coast. Where it does occur, it culminates in very old terraces that harbor a unique vegetation type dominated by the Mendocino pygmy cypress (*Hesperocyparis pigmaea*) and Bolander pine (*Pinus contorta* ssp. *bolanderi*). These terraces have remained somewhat flat for as long as a half of a

Looking across to a coastal prairie and bishop pine forest at Salt Point State Park.
John O. Sawyer

Top: Sea stack off the coast of Mendocino Village with Mendocino Coast Indian paintbrush, beach buckwheat, and seashore lupine.

Above: North Coast bluff scrub with cream cups and perennial goldfields.
TERESA SHOLARS PHOTOS

million years, creating a nutrient-poor, extremely acidic soil that has given rise to the famous stunted "pygmy" forest vegetation type. In the following sections the unique sequence of terrace vegetation is described along a transect running from the ocean to about 8 km inland.

The First Terrace

This terrace commonly occurs around 12–30 m in elevation and is constantly buffeted by windblown salts from the sea. North Coast bluff scrub, dominated by low-growing perennial shrubs, occurs along the bluff edge, blending with coastal prairie, which is dominated by perennial grasses. Further away from the ocean, scattered beach pines increase in density until they form the beach pine forest alliance. As the distance from the ocean increases, bishop pine begins to dominate and beach pine drops out, giving rise to the bishop pine forest.

Parts of the first terrace have been separated from the mainland by coastal erosion. These small islands, known as sea stacks, are owned by the Bureau of Land Management. Many of these small islands are covered by exotic species of the ice plant (Aizoaceae) family, but some stacks harbor excellent examples of North Coast bluff scrub.

It is difficult to characterize the North Coast bluff scrub because the shape, color, and overall aspect of this vegetation changes with the seasons. In early spring the low shrubs have not yet put on their new growth, so they are indistinct and only a few inches high. This allows the area to be for a short time dominated by the more ephemeral annuals like Menzies' baby blue eyes (*Nemophila menziesii*), cream cups, California goldfields (*Lasthenia californica*), woodland tidy-tips (*Layia gaillardioides*), and California buttercup (*Ranunculus californicus*). The yellow footsteps of spring sporadically dot this early-blooming community. On one site in Fort Bragg, at the Glass Beach headlands, the yellow Point Reyes blennosperma (*Blennosperma nanum* var. *robustum*) covers this unique first terrace, overlain with an old dune system. This species only occurs in two sites; here and at Point Reyes, California.

In April and May the shrubs begin their spring growth while the annuals are setting seed. The low-growing shrubs that give this diverse community its name now populate the bluffs and adjacent areas on the first terrace. Seashore lupine, seaside daisy, sea thrift, and gum plant (*Grindelia stricta*) now dominate the landscape. Also occurring are California poppy (*Eschscholzia californica*) and the rare hemiparasite Mendocino Coast Indian paintbrush (*Castilleja mendocinensis*). Inland, the California poppy is orange-flowered and generally annual. On the coast it is perennial with thicker, glaucus leaves and a yellow flower that often has an orange center. The Mendocino Coast Indian paintbrush is sporadic on the mainland but much more common on the more protected sea stacks. Other rare plants that occur here are perennial goldfields (*Lasthenia californica* ssp. *macrantha*) and, in areas of sparse cover, the diminutive short-leafed evax (*Hesperevax sparsiflora* var. *brevifolia*).

Later in the summer the flowering palette changes dramatically from the yellows of coastal gum plant to Bolander's goldenaster (*Heterotheca sessiliflora* ssp. *bolanderi*), woolly sunflower (*Eriophyllum lanatum* var. *arachnoideum*), and seaside woolly sunflower. Other species adding to the display are sea thrift, angelica (*Angelica hendersonii*), brownie thistle (*Cirsium quercetorum*), and cow parsnip (*Heracleum maximum*). This rare and endangered low shrub community occurs immediately along the ocean bluff. Its eastern limit is determined by the distribution and intensity of wind and salt spray. Coastal development and invasive plants, especially ice plants, threaten this community.

Coastal prairie forms a complex mosaic with North Coast bluff scrub, coastal seeps, and introduced perennial grassland. The prairie grasses, with their fibrous roots, have created a dark, fertile soil enriched with humus. Because this soil is so recently derived from its bedrock of greywacke sandstone, the soil texture is very sandy—a perfect habitat for Botta's pocket gophers (*Thomomys bottae*). It is said that along with the intense, salt-laden winds, the extent of the gopher population discouraged Russian settlers from agricultural pursuits along the Mendocino Coast.

Top: North Coast bluff scrub with Mendocino Coast Indian paintbrush and seashore lupine.

Above: Coastal prairie with tufted hair grass on the first terrace.

TERESA SHOLARS PHOTOS

Top: Bishop pine forest in the natural area at the Mendocino Coast Botanical Gardens.

Above: Slink pod (Scoliopus bigelovii) *occurs as an understory plant in both the bishop pine forest and redwood forest.*
TERESA SHOLARS PHOTOS

Characteristic of this community are California oat grass (*Danthonia californica*), red fescue, tufted hair grass (*Deschampsia cespitosa*), Pacific reed grass (*Calamagrostis nutkaensis*), meadow barley (*Hordeum brachyantherum*), and seaside brome (*Bromus maritimus*). In wetter microsites, patches of native Pacific wild rye (*Leymus pacificus*) and Vancouver wild rye (*L.* × *vancouverensis*) occur. Most of these sites have been highly impacted and many natives have been replaced by weedy exotics. In a few sites of low growth, the rare Blasdale bent grass (*Agrostis blasdalei*) can be found.

Introduced perennial grasses replace native coastal prairie or any native community if disturbed, plowed, and/or grazed. Most of our coastal grasslands are dominated by this vegetation type due to a long history of human disturbance. Mostly native to Europe, these exotic species are pre-adapted to grazing, trampling, and soil compaction. Common species include sweet vernal grass (*Anthoxanthum odoratum*), common velvet grass (*Holcus lanatus*), wild radish (*Raphanus sativus*), ripgut brome, wild and slender oats (*Avena fatua* and *A. barbata*), and perennial rye grass (*Lolium perenne*).

In Mendocino County the bishop pine forest occurs on this first terrace, some distance back from the ocean bluff. It also occurs on higher terraces on sandy soils or on ancient dune deposits. Bishop pine is tolerant of infertile soils, salt-laden wind, a high water table, and fire. Its serotinous cones remain closed upon seed maturity until opened by heat or age. Other species occurring with bishop pine are beach pine, coast twinberry, western sword fern (*Polystichum munitum*), and false lily-of-the-valley (*Maianthemum dilatatum*).

This forest type is in decline due to the bishop pine's short life span (80–100 years), a lack of fire, and increased coastal development. Pathogens such as western gall rust (*Endocronartium harkensii*), parasites including dwarf conifer mistletoe (*Arceuthobium littorum*), and subsequent attack by bark beetles (*Dendroctonus* spp.) have also contributed to the forest's decline. Fire suppression has caused many problems. Fire stimulates cones to open, while providing a mineral soil and full sun for optimal seedling establishment. Fire also decreases the buildup of forest pathogens.

A related closed-cone pine, Monterey pine (*Pinus radiata*), is native to just a small area on the Monterey Peninsula, but has been planted extensively outside its range. It reseeds readily on the North Coast, and can spread diseases that attack bishop pine. With the increased ornamental plantings of Monterey pine, the closed-cone pines of the coast are no longer protected by their isolation from other diseased stands.

Coastal scrub is dominated by taller shrubs than the coastal bluff scrub. It is most common on windblown slopes near the ocean, where fog is prevalent. Plants rarely grow over 1.7 m and are tolerant to salt spray. Common plants include coyote brush, bush monkeyflower (*Diplacus aurantiacus*), cow parsnip, salal, silk tassel, California blackberry, western sword fern, Wight's paintbrush (*Castilleja wightii*), and California bee plant (*Scrophularia californica*).

Second Terrace

On old dunes on the second, third, and fourth terraces, bishop pine and Douglas fir form stands with Fort Bragg manzanita (*Arctostaphylos nummularia*) in the understory. Golden chinquapin (*Chrysolepis chrysophylla*) also occurs on these more porous soils. Occasionally on the second terrace, but most often on the slopes and alluvial sites away from the immediate areas of salt spray, is the redwood forest alliance. This forest is a mosaic of tree species that depends on exposure to ocean winds, soil quality, moisture, aspect, temperature, and successional stage. On better soils, specifically on slopes and alluvial plains, redwood is dominant. Closer to the coast, grand fir, which is tolerant of windblown salts and seeds best in shady sites, is most common. In warmer and drier sites, Douglas fir is characteristic while in

Top: Redwood forests occur on the second terrace, ridges, slopes, and alluvial flood plains. Robert Wick

Inset: The mycorrhizal winter chanterelle (Craterellus tubaeformis) *is common in the redwood forest.* Teresa Sholars

Top: Hiking in Jug Handle State Natural Reserve in Mendocino County. Podolized soils nurture a "pygmy" forest with rare plants like Mendocino pygmy cypress and pygmy manzanita (Arctostaphylos nummularia *ssp.* mendocinoensis). MICHAEL E. KAUFFMANN

Above: Mendocino pygmy cypress woodland. TERESA SHOLARS

shadier and wetter sites, western hemlock is more common. There are a few locations along the immediate coast where one can see Sitka spruce, especially in the vicinity of Caspar Creek which is near its southern limit. Occurring in the understory throughout this forest are tanoak, western sword fern, red huckleberry (*Vaccinium parvifolium*), evergreen huckleberry, rhododendron (*Rhododendron macrophyllum*), salal, and redwood sorrel (*Oxalis oregana*). This conifer forest hosts a variety of large, obligate, mycorrhizal mushroom species that makes this area a mecca for mushroom hunters. Symbiotic associations form between the roots of most woody plant species and fungi, where carbon flows to the fungus from plant photosynthates, and inorganic nutrients move to the plant from the fungus. This association greatly increases the root absorptive area, giving the symbionts a huge competitive advantage. Most of the large mushrooms that occur in the forest are dependent on the trees for their sources of energy.

THIRD, FOURTH, AND FIFTH TERRACES

Mendocino pygmy cypress woodland occurs only in a few sites along the Mendocino Coast, on the third, fourth, or fifth terraces. The soils here are 1,000 times more acidic than soil found in the redwood forest. Heavy winter rains have leached iron, clay, and soil nutrients from the surface of the ground and washed them down to the subsoil. The iron, soluble under acidic conditions, combines with eroding bedrock (subsoil) to form an iron-concreted hardpan 25 cm below the surface in many sites. Extremely acidic conditions and poor soil fertility, coupled with shallow hardpan formation and wet soil (bog conditions), contribute to the stunted, sparse growth of the pygmy forest. When soil is wet, water replaces oxygen in the soil's porous spaces, and because of this, plants of the pygmy forest must tolerate low soil oxygen levels. Since roots take in oxygen and give off carbon dioxide, most plants are not able to live in soils that are too wet. For all of the reasons above, one hundred year old trees may only reach a meter in height. The ground and trees are covered by many species of lichens due to the availability of light, the suitably moist climate, and the clean air. These soil lichens help prevent erosion caused by water movement. Characteristic plants of the pygmy forest include Mendocino pygmy cypress, Bolander pine, Fort Bragg manzanita, and California sedge (*Carex californica*). Many of the redwood understory species occur here in a stunted condition, such as rhododendron, evergreen huckleberry, and salal. The pygmy forest has been subjected to increased development, and myriad new roads and drainage ditches cutting through the landscape result in increased erosion and transport of nutrients into the pygmy forest, creating a taller and much different forest.

Estuaries

Estuaries have suffered greatly in northern California, as in the rest of the state, from human impacts. The largest estuaries in the region are Humboldt Bay and the mouth of the Eel River in Humboldt County. Both of these estuaries were extensively diked to convert salt marsh to agriculture in the early 1900s and now have only 10% of their original salt marsh acreage remaining. A number of smaller bays and estuaries occur throughout the region including Bodega Bay, Tomales Bay, Drakes Estero, and the mouths of rivers to the north and south, including Smith River, Redwood Creek, Mad River, Mattole River, Ten Mile River, Big River, Little River, Albion River, and Russian River. There are also intermittently breached lagoons in Humboldt and Del Norte counties such as Lakes Earl and Tolowa, Big Lagoon, and Stone Lagoon. These estuaries, bays, and lagoons all play vital roles in both oceanic and terrestrial systems. They are productive systems that serve as rearing grounds for fish and shellfish, and provide important stopover and nesting habitat for migratory birds.

The vegetation of estuaries is structured along gradients of salinity and tidal inundation/exposure. Eelgrass (*Zostera marina*) beds grow at the lowest intertidal zones and in areas of total submergence. These dense stands of grass-like aquatic plants are restricted to certain bottom sediment sizes and can be negatively affected by turbidity or poor water quality. Eelgrass is the primary dietary component of black brant (*Branta bernicla nigricans*), a goose that migrates along the west coast in spring. It also forms the base of an important detritus-based estuarine food web.

Higher in the intertidal zone are salt

Top: Tidal flats and eelgrass beds in Humboldt Bay National Wildlife Refuge.
Gary Todoroff

Above: A river otter (Lontra canadensis) *surfaces in Big River estuary at sunset.*
Andrea Pickart

Top: Salt grass flats in winter at Humboldt Bay. The interspersed pickleweed turns red as it enters dormancy.

Above: The hemiparasitic Point Reyes bird's-beak grows in the high marsh, where it has been shown to ameliorate harsh conditions. Andrea Pickart photos

marshes, which are underappreciated landscapes of great beauty. Meandering tidal creeks form a vast network through these marshes, delivering nutrient-rich tidal waters. The North Coast's salt marshes have a flora distinct from those to the north, which are less saline and dominated by Lyngbye's sedge (*Carex lyngbyei*). California cordgrass (*Spartina foliosa*) populates the low marsh only at the southern end of the region. North of Bodega Bay, there is no ecological analog for this species, and low marsh is characterized by pickleweed (*Sarcocornia pacifica*) mats, while high marsh supports salt grass (*Distichlis spicata*) flats.

The highest marshes are quite diverse. A close look is required to appreciate this diversity, as most species are small, low-growing, and have inconspicuous flowers. However, there are charismatic salt marsh plants as well. In a good year the annual, rare hemiparasite Humboldt Bay owl's clover (*Castilleja ambigua* ssp. *humboldtiensis*) forms breathtaking expanses of fuchsia on the high marsh and in relatively brackish areas. The related rare hemiparasite Point Reyes bird's-beak (*Cordylanthus maritimus* ssp. *palustris*) has been shown to reduce stressful conditions on the marsh, thereby increasing native species diversity. A true parasite, goldenthread (*Cuscuta salina* var. *major*), is also found on North Coast salt marshes. Its tangle of bright orange stems can be found among its hosts, pickleweed and marsh jaumea (*Jaumea carnosa*). Its flowers are relatively conspicuous, and when it twines around the stems of pickleweed, a casual observer can mistake the flowers for those of the pickleweed. Salt marsh dodder, like the hemiparasites, helps to increase species diversity on the salt marsh by, in the case of dodder, suppressing its dominant host. Other species on the marsh include slender arrowgrass (*Triglochin concinna*), the rare western sand spurrey (*Spergularia canadensis* var. *occidentalis*), salt marsh plantain (*Plantago maritima*), sea lavender (*Limonium californicum*), marsh jaumea, and gum plant.

Humboldt Bay and Eel River estu-

aries have been relentlessly invaded by the introduced South American Chilean cordgrass (*Spartina densiflora*), which forms its own alliance known as Chilean cordgrass marshes. Although Chilean cordgrass has a much higher above-ground biomass than native marsh species, the net productivity of invaded marshes is lower, in part because of its suppression of the algal community that is an important part of salt marsh productivity. Invasion has also been found to reduce invertebrate richness, causing a shift from marine to terrestrial invertebrates. Eradication of this species is now underway at Humboldt Bay. San Francisco Bay was similarly invaded by smooth cordgrass (*S. alterniflora*), introduced in the 1970s. The species quickly hybridized with the native California cordgrass, which ranges from Tomales Bay south into Mexico. Smooth cordgrass hybrids displace mud flats, alter sedimentation patterns, and cause a shift in the functional role of benthic invertebrates. A 13-year multi-agency eradication program spearheaded by the California State Coastal Conservancy has reduced the invasion of this and other invasive *Spartina* species by 95%.

Salt marsh plants are uniquely adapted to tolerate the biological extremes posed by daily flooding and exposure, and the toxicity of salts in seawater. Through a variety of mechanisms, including salt secretion or compartmentalization, and translocation of oxygen, these plants overcome anoxia and soil salinity. Salt marshes are areas of sediment accretion in the estuary, and are normally able to keep pace with sea level rise. However, their ability to keep up with accelerated, climate-change-induced sea level rise will vary from place to place, depending on levels of suspended sediment, rates of inter-seismic and/or co-seismic subsidence, and other factors. Certainly, they must have accommodation space in order to migrate inland with rising sea levels. In our larger estuaries, dikes act as barriers to migration.

In areas of greater freshwater input,

Top: Invaded Chilean cordgrass marshes, Indian Island, Humboldt Bay.

Above: High marsh (salt grass flats alliance) with Humboldt Bay owl's clover.

ANDREA PICKART PHOTOS

Top: Tufted hair grass meadows in a brackish marsh, Humboldt Bay.

Above: Bur-reed flowers in a brackish marsh.

Andrea Pickart photos

either farther upstream at the edge of terrestrial ecosystems or lining infrequently breached lagoons, salt marshes give way to brackish marshes. Tidal brackish areas may support salt marsh bulrush (*Bolboschoenus maritimus*) marshes, salt rush (*Juncus lescurii*) swales, California bulrush (*Schoenoplectus californicus*) marshes, or tufted hair grass meadows. Often the most diverse stands of vegetation are found at the ecotone between salt and freshwater marshes. The temporally variable influences of tide, rainfall, and fresh groundwater allow for the co-occurrence of species from the full freshwater–salt-water gradient. Found in a narrow band lining the upper edge of the salt marsh are bur-reed (*Sparganium erectum* ssp. *stoloniferum*), western lilaeopsis (*Lilaeopsis occidentalis*), sea milkwort (*Glaux maritima*), mulefat (*Baccharis salicifolia*), rare seacoast angelica (*Angelica lucida*), and spear oracle (*Atriplex patula*).

Riparian and Freshwater Wetlands

The only forests on the North Coast dominated by deciduous trees are those found in freshwater swamps and along the banks of creeks and rivers. Red alder, willows (*Salix* spp.), cascara buckthorn, Pacific red elderberry (*Sambucus racemosa* var. *racemosa*), and Oregon crabapple (*Malus fusca*) dominate on the immediate coast, and bigleaf maple (*Acer macrophyllum*) and Oregon ash (*Fraxinus latifolia*) occur a bit more inland. Understory species vary considerably and overlap with the freshwater wetland species listed below. Typical riparian understory species include pink-flowering currant, spikenard (*Aralia californica*), western sweet coltsfoot (*Petasites frigidus* var. *palmatus*), coast hedge nettle, coastal wood fern (*Dryopteris arguta*), five finger fern (*Adiantum aleuticum*), and the horsetails (*Equisetum telmateia*,

E. arvense). North Coast deciduous plants lose their leaves in late fall, letting light and warmth in during winter months while a new set of leaves shades the rivers in the summer. This keeps the water temperature at levels that provide optimal conditions for fish such as steelhead trout (*Oncorhynchus mykiss*) and coho salmon (*Oncorhynchus kisutch*). Riparian plant species are tolerant of flooding and variable levels of soil oxygen. Flooding suspends and deposits fine sediments on the surrounding land, creating broad, fertile floodplains.

Freshwater marshes occur throughout the North Coast, wherever the water table is at least seasonally high or there is surface water accumulation. Common along the coast are small wetlands dominated by sedges (*Carex* spp.) and rushes (*Juncus* spp.), which form mosaics with forests and grasslands. Other plants associated with these small wetlands

Top: Red alder forest with an understory of western sword fern and slough sedge along Ossagon Creek, Redwood National and State Parks.

Above: Riparian vegetation with flowering spikenard and five finger fern in the Sinkyone Wilderness State Park.

Andrea Pickart photos

Top: A seasonal freshwater wetland supports slough sedge swards with flowering hedge nettle at Point Reyes National Seashore.

Above: Duckweed and mosquito fern float in an open water wetland. Andrea Pickart photos

are deer fern (*Blechnum spicant*), lady fern (*Athyrium filix-femina* var. *cyclosorum*), fringed false-hellebore (*Veratrum fimbriatum*), and skunk cabbage (*Lysichiton americanus*). In areas where standing water persists, emergent and floating herbs colonize, such as water-parsley (*Oenanthe sarmentosa*), common monkey flower (*Mimulus guttatus*), tall mannagrass (*Glyceria striata*), duckweed (*Lemna* spp.), mosquito fern (*Azolla* spp.), yellow pond-lily (*Nuphar lutea* ssp. *polysepala*), and pondweeds (*Potamogeton* spp.).

Among the most picturesque wetlands of the North Coast are the coastal lagoons. Infrequently breached, they support brackish to freshwater vegetation along their margins, including hardstem bulrush (*Schoenoplectus acutus*) marsh and cattail (*Typha* [*angustifolia, domingensis, latifolia*]) marshes. Lakes Earl and Tolowa, which actually comprise one body, are the largest of the lagoons, with smaller water bodies at Big Lagoon, Stone Lagoon, and Freshwater Lagoon. Dry Lagoon no longer supports permanent open water, filling with a rich assortment of marsh vegetation during the summer months. The Tolowa Dunes are also home to relatively large inland water bodies, such as Dead Lake, which are associated with stabilized, forested dunes.

Wet pastures, or agricultural wetlands, are prominent in large estuaries such as Humboldt Bay located behind dikes in areas that once supported tidal marsh. The vegetation in these areas features a continuum of native to introduced grasses and forbs, with species composition structured by water regime and salinity. The highest and driest sites are typically grazed and support two semi-natural alliances: perennial rye grass fields and common velvet grass–sweet vernal grass (*Holcus lanatus–Anthoxanthum odoratum*) meadows. Localized stands of native grasses, including red fescue, creeping rye grass (*Leymus triticoides*), and water foxtail (*Alopecurus geniculatus, A. aequalis*), can be found interspersed with these introduced species, especially in moist microsites.

As pastures get wetter, they grade into native wetland alliances, including Pacific silverweed marshes, salt rush swales, water-parsley marshes, and pale spike rush (*Eleocharis macrostachya*) marshes. Saline areas, especially disturbed sites, are characterized by ephemeral fields of fat hen–brass buttons (*Atriplex prostrata–Cotula coronopifolia*) semi-natural stands.

Unique wetlands, known generally as peatlands, are found in only a few places along the North Coast. Often called fens or bogs, they are more extensive at boreal latitudes. Inglenook fen, nestled in the Ten Mile Dunes, is fed by Fen Creek. Created thousands of years ago, it has persisted through ongoing shoreline changes and has accumulated up to 9 m of peat. Big Lagoon Bog in Humboldt County has accumulations of up to 3.6 m of peat. Terminology of peatlands is often ambiguous and colloquial, but both of these sites would be properly classified as fens based on their groundwater inputs. Both sites are rich in rare and unusual taxa, including swamp harebell (*Campanula*

Top: Water-parsley marsh in a freshwater wetland at Humboldt Bay National Wildlife Refuge.

Middle: Hardstem bulrush marsh can be seen in the background behind dune vegetation at Dry Lagoon. This lagoon no longer supports open water during summer months.

Right: Inglenook fen-carr abuts Sandhill Lake, MacKerricher State Park.

ANDREA PICKART PHOTOS

Top, a salt marsh experiencing a normal tide, and above, a "King Tide" at the same location. With sea level rise and climate change, coastal salt marshes will be subjected to extreme tides more often. ANDREA PICKART PHOTOS

californica), coast lily (*Lilium maritimum*), bog clubmoss (*Lycopodiella inundata*), Labrador tea (*Ledum* × *columbianum*), and bog bean (*Menyanthes trifoliata*). Some of these species may be Pleistocene relicts, persisting in these cool, moist microsites. Both peatlands are surrounded by a "fen-carr" or "carr," terms for a fen that has overgrown with shrubs and trees, in this case red alder, willows, coast twinberry, and wax myrtle. Due to their extreme sensitivity, these sites are not open to the public. Inglenook fen abuts a large freshwater pond known as Sandhill Lake, a body of open water edged with bulrush, yellow pond-lily, marsh cinquefoil (*Comarum palustre*), and, climbing the surrounding dunes, scouring rush (*Equisetum hyemale* ssp. *affine*).

Looking Ahead on the Northern California Coast

After more than a century of focus on development pressures joined by invasive species threats, the spotlight of conservation concern for northern California vegetation has shifted to climate change. Rising seas, warming temperatures, and changes in precipitation, storm frequency, and storm intensity are combining to bring unprecedented rates of change for coastal ecosystems. The mean sea level in northern California is predicted to rise by up to 0.5 m by 2050 and 1.5 m by 2100. The frequency of extreme sea levels is also predicted to increase due to storm surge and wave run-up associated with extreme events. Although inter-seismic land-level change may slow the rate of sea level rise in some parts of the Cascadia subduction zone, it has been shown to amplify it in other places, such as Humboldt Bay. Rupture of the subduction zone, which last occurred in 1700, could result in instantaneous subsidence of 1–2 m. Whereas these episodic, catastrophic events are not new to our North Coast ecosystems, in the past, equilibration has been able to occur with respect to relatively static climatic conditions.

The vegetation of the North Coast will be particularly susceptible to sea level rise. The cyclic ocean-warming El Niño events that characterize our coast can exacerbate sea level rise by as much as 30 cm, and extreme El Niños are predicted to become more frequent. Increased wave energy during storm events is expected to cause the retreat of

bluffs and, where accommodation space exists, the landward migration of dunes. Salt marshes may be able to keep pace with sea level rise where there is sufficient space for them to migrate, but many of our estuaries are bounded by dikes or other structures. The fate of both sand dunes and salt marshes depends on sediment supply, a difficult variable to predict. Sadly, coastal bluffs, unlike salt marshes and dunes, are fixed geomorphic features and face permanent losses.

Land managers, scientists, and conservationists are working to find pragmatic, adaptive responses to climate change. In addition to direct impacts on vegetation—such as range shifts caused by changing temperatures and precipitation—coastal flora will be subject to pressures from new or accelerated plant and pathogen invasions, decoupling of plant-animal interactions, and changes to disturbance regimes that have long underpinned native plant communities. Amidst much uncertainty, conservation scientists agree on the importance of preserving biological diversity and ecosystem function. Ecosystem, species, and genetic diversity increase adaptive capacity, and the North Coast faces the challenge of climate change bolstered by high levels of biodiversity. Restoration of ecosystems will focus more on restoring function and process and less on preserving historic species assemblages. Those who love California's unique flora and vegetation will need to advocate for their continued protection.

Andrea J. Pickart
Teresa A. Sholars

Sitka spruce on Pewetole Island offshore of Humboldt County. This island and a much smaller seastack just south of it are the furthest south locations where Sitka spruce can be found on offshore rocks.
Robert Wick

CHAPTER TWO

Northern California Coast Ranges and Interior Ranges

California's North Coast is famous for its majestic redwood (*Sequoia sempervirens*) forests, which contain the tallest and some of the largest trees in the world. Along the coast, the redwood forests are some of the most productive forestlands in the world. Inland from the redwoods, a variety of forest types mix with chaparral, grasslands, and woodlands at lower elevations while fir forests clothe the higher elevations. The area covered in this chapter extends from the southern limit of the Klamath Region south to Marin County and inland to the Sacramento Valley (Hickman 1993). The immediate coast is covered in Chapter One, Northern California Coast.

The region's climate is overall Mediterranean, characterized by dry, hot summers and cool, wet winters. Being near the coast, conditions are moderated in the summer by fog and cool temperatures. During winter, snow is only occasional and fleeting at the lowest elevations, while at higher elevations the mountain ranges maintain snow well into the summer. The highlands near Crescent City and those surrounding Cape Mendocino are some of the wettest places in the state. The eastern side

Above: Douglas fir forest mixes with grasslands and oak woodlands in the mid-elevations of the North Coast in coastal Humboldt County.
JOHN O. SAWYER

Opposite: Pacific rhododendron (Rhododendron macrophyllum) *in bloom in a redwood forest along the Damnation Creek Trail in Del Norte Coast Redwoods State Park.*
LARRY ULRICH

Top: This extensive Douglas fir–tanoak forest occurs where the North Coast meets the Klamath Mountains. Robert Wick

Above: The southeastern foothills of the inner North Coast in Napa County exhibit a variety of vegetation types, including chaparral and oak and foothill pine woodlands on the distant steeper slopes, and serpentine grassland on the closer gently rolling slopes. Julie M. Evens

lies outside of the summer fog belt and summers here are hot. It does lie within the rain shadow of the Yolla Bolly Mountains and the North Coast Ranges, making it the driest region in northwestern California.

Broad Vegetation Pattern

The vegetation pattern changes over three gradients: traveling inland from the ocean, increasing in elevation, and going from north to south. Along the coast and select river valleys, forests are often dominated by redwoods, whose noncontiguous range extends from southwestern Oregon to Monterey County. Tanoak (*Notholithocarpus densiflorus*) has a similar range but tends to be more common slightly inland from the coast in this region. Douglas fir (*Pseudotsuga menziesii*), which is common throughout the North Coast, mixes with tanoak, creating Douglas fir–tanoak forests, an extensive forest alliance inland from the immediate coast. In the sunny and warmer interior lands east of the crest, Douglas fir forests are extensive at low elevations. Chaparral, grasslands, and woodlands become more common in the interior lowlands of the southernmost parts of the region.

In the higher elevations, extensive forests dominated by white fir (*Abies concolor*) distinguish the beginnings of the montane coniferous forests. While white fir's presence indicates a shift into low montane elevation, the presence of Shasta red fir (*A. magnifica* var. *shastensis*) indicates passage into the upper montane. Shasta red fir and high-elevation pines characterize the subalpine belt, which is restricted to the highest peaks and ridges in the South Yolla Bolly Mountains around Mount Linn. Here, in addition to scattered western white pine (*Pinus monticola*), one population of foxtail pines (*P. balfouriana*) still persists.

The eastern foothills support forests containing Douglas fir, foothill pine (*P. sabiniana*), and ponderosa pine (*P. ponderosa*), as well as chaparral, grasslands, and oak (*Quercus* spp.) woodlands. To the south, both conifer and hardwood forests are important in the wetter coastal ranges; chaparral and woodlands, however, replace forests in the drier inland ranges.

As in the Klamath Mountains region, serpentine outcrops interrupt broad vegetation patterns found on soils with high fertility and water holding capacity. Serpentine outcrops are found throughout the North Coast, with the larger ones occurring in the mountains and foothills in the southern parts of the

region. These serpentine outcrops, associated with the Cascadia subduction zone, separate the more predictably productive rock types of the Great Valley sequence to the east, from those of the Franciscan Complex and volcanics to the west. The Franciscan Complex includes mélanges of sheared and crushed fine-grained mudstones and shale containing scattered rock outcrops of chert, greenstone, and limestone. The weathered surface layers of the mélange creep down-slope and create earth flows that result in a distinctive hummocky terrain and mosaic of forests, woodlands, and prairies that characterize the ranchlands inland from the redwoods.

Watersheds of the North Coast

The North Coast Ranges extend from coastal Oregon south to San Francisco, and are inland from the immediate coast and west of the Klamath Region and the Sacramento Valley. The broad forest pattern is best appreciated by considering separately the lands north of the Eel River, those of the Eel River, those of the

The only population of foxtail pines in the North Coast occurs in a small grove on Mount Linn, the highest peak in the region.
Michael E. Kauffmann

Mixed conifers of western hemlock and redwood comprise the redwood forests in Prairie Creek Redwoods State Park. Jeff Bisbee

Russian River and Napa Valley, and those on the east side of the Coast Ranges.

North of the Eel River

The lands north of the Eel River, near Eureka, support the state's temperate rain forests with redwood blanketing the low mountains along the coast surrounding Humboldt Bay, and northward to Mad River, Redwood Creek, Klamath River, and Smith River. Forest canopies are a mixture of conifers: redwood, Douglas fir, Port Orford cedar (*Chamaecyparis lawsoniana*), western redcedar (*Thuja plicata*), grand fir (*Abies grandis*), and western hemlock (*Tsuga heterophylla*). Hardwoods of the region include California bay (*Umbellularia californica*) and madrone (*Arbutus menziesii*), with tanoak trees being uncommon. The understories are filled with ferns, especially western sword fern (*Polystichum munitum*). Sitka spruce (*Picea sitchensis*) joins redwood on the lower slopes, stream terraces, and alluvial flats. These associations are easily seen in the Arcata Community Forest, Redwood National Park, Del Norte, Prairie Creek, and Jedediah Smith Redwoods state parks, and Headwaters Forest Reserve.

As redwood trees age, they develop complex canopy crowns due to wind events and the damage caused by nearby falling trees and branches. In response to injury, trees sprout new leaders while injured ones die or break off. As a result, the crowns of large, old trees become a complicated mass of many separate leaders, some living and others dead. In these reiterations and limbs, detritus collects and, if large enough, becomes a medium for plant growth. As in the temperate coastal rainforests to the north, epiphytic lichens, mosses, and liverworts grow in these mats, as do ferns such as leather fern (*Polypodium*

scouleri), and other forest shrubs like salal (*Gaultheria shallon*) and evergreen huckleberry (*Vaccinium ovatum*). Even small western hemlock and Sitka spruce trees can be found in the canopies under proper conditions.

The Eel River

Redwood forests in the Eel River watershed differ in several ways from those to the north. Here the common associates include California bay, Douglas fir, madrone, and tanoak, while Sitka spruce, western redcedar, and western hemlock are absent. Forests on terraces along major tributaries are nearly pure redwood; slope forests are a mixture of Douglas fir, redwood, the occasional grand fir, and hardwoods—especially tanoak. Slope trees are short, and the forest canopy is more open. In contrast, trees on the alluvial flats are tall, including many of the tallest trees in the world, and the forest canopy is closed. Large, old tree crowns are less complex and epiphytic mats are smaller. Slope redwood forests give way to Douglas fir–tanoak forests on drier sites. The transition between two forest types is sometimes gradual. Other times, however, it is abrupt, where slopes act as sources of seepage water. These

Top: Complex canopy of old redwood trees in Redwood National Park.
Steve Norman

Above: Leather fern growing on a detritus mat accumulated in the canopy of an old redwood tree.
Steve Sillett

Top: In the King Range, where redwoods are nearly absent, Douglas fir forests are extensive with an outlying population of sugar pine decorating the south-facing slopes on the highest peaks.

Above: Redwood and tanoak riparian along the Eel River.
Robert Wick photos

redwood and Douglas fir forests are the two major forest types in Humboldt Redwoods State Park and other state parks of Humboldt and northern Mendocino counties.

From the mouth of Eel River, around Cape Mendocino and south to Cape Vizcaino, redwood forests lie inland from the coast. This is the "Lost Coast," a 96-kilometer roadless stretch of wild coastline encompassing the King Range Wilderness (within the King Range National Conservation Area) as well as Sinkyone Wilderness State Park. Here, Douglas fir–tanoak forests are extensive and contain California bay, canyon live oak (*Quercus chrysolepis*), and madrone. Redwood forests again approach the coastline east of the coastal terraces south of Cape Vizcaino and continue to Sonoma County.

Elsewhere, much of the outer Coast Range land is privately owned timberlands. The Jackson State Forest, bisected by Highway 20 from Willits to Fort Bragg, is the largest of the California Department of Forestry's eight demonstration forests. These are managed for recreation and to demonstrate proper harvesting, watershed management, and restoration techniques to private landowners. Jackson State Forest is adjacent

to Big River and Mendocino Headlands state parks, and University of California's Angelo Coast Range Reserve, acquired first by The Nature Conservancy. West of Anderson Valley, Navarro Redwoods State Park follows the Navarro River to its mouth.

Inland, the Bald Hills prairies are extensive along the rounded ridges of the outer Coast Range and the western slopes of the inner Coast Range. Oak woodlands surround the prairies; Douglas fir forests inhabit north-facing slopes. Knobcone pine (*Pinus attenuata*) forests are scattered and Douglas fir commonly mingles with canyon live oak, foothill pine, and ponderosa pine. Steep, rocky slopes are covered with interior live oak (*Quercus wislizeni*) and California bay forests. Serpentines in Eden Valley, east of Covelo, support a large Sargent cypress (*Hesperocyparis sargentii*) grove. Small serpentine patches support stands of incense cedar (*Calocedrus decurrens*).

In the woodlands, Oregon white oak (*Q. garryana*) is the most familiar and widespread oak and often the only one present. California black oak (*Q. kelloggii*) occurs occasionally and blue oak (*Q. douglasii*) grows south of Round Valley. Open stands of canyon live oak and interior live oak dominate rocky sites. Valley oak (*Q. lobata*) inhabits deep, loamy soils along the main Eel River. Understories often contain California fescue (*Festuca californica*) or tall oat grass (*Arrhenatherum elatius*). Historically, species composition of the prairies has varied according to local ranch grazing practices. Annual dogtail (*Cynosurus echinatus*), soft brome (*Bromus hordeaceus*), California oat grass (*Danthonia californica*), longbeak stork's bill (*Erodium botrys*), and rose clover (*Trifolium hirtum*) are common. Degraded sites support Klamath weed (*Hypericum perforatum*) and yellow star-thistle (*Centaurea solstitialis*). Tracts of chamise

Prairies with post-fire bloom of riverbank lupine (Lupinus rivularis) *surrounded by Oregon white oak woodlands; Bald Hills, Redwood National Park.* Andrea Pickart

Top: The northern range extension for Sargent cypress occurs on Red Mountain, here with sugar pine high above the Eel River.
Michael E. Kauffmann

Above: Ponderosa pine and incense cedar after a fire in the BLM Red Mountain unit, Mendocino County.
Robert Wick

(*Adenostoma fasciculatum*), hoary manzanita (*Arctostaphylos canescens*), or wedge leaf ceanothus (*Ceanothus cuneatus*) chaparral enrich the mosaic.

At montane elevations in the inner Coast Range, conifers become more common with increasing elevation, as Douglas fir forests give way to ponderosa pine–incense cedar forests and then to white fir forests. Knobcone pine forests become small and uncommon. Local stands of incense cedar occur on ridges and at meadow edges. Shasta red fir forests clothe the highest ridges and peaks. Western juniper (*Juniperus occidentalis*) woodlands grow on sandstone and shale in the Yolla Bolly Mountains and scattered ridgelines westward to Humboldt County. Serpentine outcrops at Red Mountain and Mount Lassics, west of the major mountain crest, support open Jeffrey pine (*Pinus jeffreyi*) forests. Near the Humboldt–Mendocino county line, the northern-most stands of Sargent cypress grow on Red Mountain where they associate with sugar pine (*P. lambertiana*), incense cedar, and Jeffrey pine.

The Russian River and Napa Valley

The vegetation in the southernmost North Coast Ranges differs in several ways from the more-heavily forested north. Hardwood forests and woodlands

are extensive, and chaparral and grasslands are widespread. Much of the region is committed to agriculture, especially vineyards. Santa Rosa is the largest of the many towns in southern Mendocino, Napa, and Sonoma counties. Resort towns surrounding Clear Lake in Lake County push chaparral and woodlands up into the surrounding hills. Unlike the northern counties, most of the rural land is privately owned.

From Hull Mountain the Russian River flows southwesterly between the outer Coast Range and the Mayacamas Mountains of the inner Coast Range. It runs through Potter Valley, Ukiah Valley, and onto the Santa Rosa Plain before passing south of the high hills surrounding Rocky Mountain, to its mouth at Jenner. The lower slopes support the transition from redwood forests to both Douglas fir–tanoak and hardwood forests. Depending on elevation and aspect, forests of California bay, coast live oak (*Quercus agrifolia*), and interior live oak mix with chaparral alliances such as scrub oak (*Q. berberidifolia*), common whiteleaf manzanita (*Arctostaphylos manzanita*), and whiteleaf manzanita (*A. viscida*). The hilltops and inland slopes support a mosaic of hardwoods, Oregon white oak woodlands, and prairies. This is the vegetation mosaic of Armstrong Redwoods State Natural Reserve, Austin Creek State Recreation Area, Warm Springs Recreation Area at Lake Sonoma, and Montgomery Woods State Natural Reserve.

Top: Coast live oak on the top of a slope at Austin Creek State Recreation Area in Sonoma County.
Todd Keeler-Wolf

Above: Hoary manzanita chaparral in the Mayacamas Mountains.
Shanti Edwards

Top: Whiteleaf manzanita on serpentine at The Geysers in Sonoma County.
TODD KEELER-WOLF

Above: The California nutmeg is found in the mix-evergreen forest understory, here in Robert Louis Stevenson State Park.
MICHAEL E. KAUFFMANN

Top right: Boggs Lake Ecological Reserve, a large vernal pool with several associated rare species near Cobb Mountain, Lake County.
TODD KEELER-WOLF

The Cedars, west of Sonoma, offer a strikingly different and unique landscape (Raiche 2009). This large serpentine outcrop has extensive woodlands of Sargent cypress and serpentine chaparral of whiteleaf manzanita, leather oak (*Q. durata*), Jepson ceanothus (*Ceanothus jepsonii*), wedge leaf ceanothus, and a rare manzanita, called The Cedars manzanita (*Arctostaphylos bakeri* ssp. *sublaevis*). Barrens, talus, and bare rock support only scattered perennials, including endemics such as Hoffman's jewelflower (*Streptanthus glandulosus* ssp. *hoffmanii*) and disjuncts such as serpentine phacelia (*Phacelia corymbosa*), which also grows in the Klamath Mountains and the Sierra Nevada.

Cow Mountain, Cobb Mountain, Mount Saint Helena, and other peaks of the Mayacamas Mountains lie west of Clear Lake and Napa Valley. Highlands between Cobb Mountain and Saint Helena include the extensive Douglas fir–ponderosa pine forests of Boggs Mountain Demonstration State Forest, and a vernal pool with Boggs Lake hedge hyssop (*Gratiola heterosepala*) and slender orcutt grass (*Orcuttia tenuis*) at Boggs Lake Ecological Reserve. Mount Saint Helena and the surrounding Robert Louis Stevenson State Park, nearby Bothe-Napa Valley State Park, Hood Mountain Regional Park, and Sugarloaf Ridge State Park offer mixed evergreen forests with California bay, Douglas fir, madrone, live oaks, tanoak, groves of redwood, and even the elusive California nutmeg (*Torreya californica*) in the understory. Chaparral here includes common whiteleaf manzanita, hoary manzanita, Stanford's manzanita (*A. stanfordiana*), chamise, and wavy leaf ceanothus (*Ceanothus foliosus*), and mixes with upland annual grasslands. Annadel State Park in the lower-elevation Sonoma Mountains contains extensive mixed woodlands of California black oak,

California bay, coast live oak, Douglas fir, madrone, and Oregon white oak. Grasslands have squirreltail (*Elymus elymoides*) and foothill (*Nassella lepida*), as well as nodding (*N. cernua*) and purple (*N. pulchra*) needle grasses.

The inner-most ridges of the inner Coast Range, from Walker Ridge east of Clear Lake, south to Mount Vacaville, and east of the Napa Valley, are largely untrammeled by humans. Clear Lake occupies a trough between Cow Mountain and Walker Ridge. The southeast flowing Cache Creek begins on the slopes of Snow Mountain and, flowing southeast, picks up water from Clear Lake, after which it ends its journey at the Sacramento River. The ridge to the south includes Baldy Mountain, Blue Ridge, Berryessa Peak, and Mount Vaca, and its complex geology is one of serpentine, sedimentary, and volcanic substrates. Fire return intervals are much shorter here than others in this region.

Chaparral clothes the steeper slopes, where vegetational change on and off serpentine is striking: leather oak, Jepson ceanothus, hoary coffee berry (*Frangula californica* ssp. *tomentella*), and whiteleaf manzanita on serpentine; birch leaf mountain mahogany (*Cercocarpus montanus*), California buckeye (*Aesculus*

Top: Serpentine outcrop at The Cedars supports the endemic The Cedars manzanita as well as stands of Sargent cypress.
Jeff Bisbee

Above: Redbud in bloom along Cache Creek.
Andrea Pickart

californica), interior live oak, scrub oak, and toyon (*Heteromeles arbutifolia*) off serpentine. Groves of McNab cypress (*Hesperocyparis macnabiana*) or Sargent cypress, some often quite extensive, signal the existence of serpentine soils. Walker Ridge, with frequent fire return intervals, is believed to nurture the most extensive stand of McNab cypress anywhere. Pure or mixed chaparral of chamise, wedge leaf ceanothus, or whiteleaf manzanita exist on and off serpentine. Individual trees or groves of foothill pine grow throughout.

Pure or mixed woodlands with California black oak, blue oak, and interior live oak are widespread, mixing with grasslands in the lower slopes and valley bottoms. On serpentine are squirreltail, curly blue grass (*Poa secunda*), and California melic grass (*Melica californica*). Off serpentine are non-native cheat grass (*Bromus tectorum*), foxtail chess (*B. madritensis*), medusa head (*Taeniatherum caput-medusae*), and wild oats (*Avena barbata, A. fatua*). Foothill pine and knobcone pine stands are scattered throughout the complex vegetation mosaic. These are the major types within the Bureau of Land Management's Cache Creek Natural Area, including Cache Creek Wilderness and the Cedar Roughs Wilderness, the state's Knoxville Wildlife Area, and University of California's Mclaughlin, Quail Ridge, and Stebbins Cold Canyon natural reserves.

Similarly, substrate defines the

Top: Birch leaf mountain mahogany chaparral near Pillsbury Lake, Lake County.
Jim Xerogeanes

Middle: Wedge leaf ceanothus helps define the chaparral of the inner North Coast.
John O. Sawyer

Left: Walker Ridge has a mosaic of chaparral plants, including whiteleaf manzanita, and McNab cypress, which depend on fire to regenerate over time.
Michael E. Kauffmann

Above: Sargent cypress and the endemic Mt. Tamalpais manzanita grow in serpentine on and around Mount Tamalpais State Park.
Jeff Bisbee

Left: Scattered foothill pine mixed with chaparral along Cache Creek downstream from Clear Lake.
Robert Wick

variation in chaparral, grassland, and forested vegetation at Mount Tamalpais, a beautiful mountain with a commanding view of Marin County and northern San Francisco Bay east of Bolinas Bay. Stands of endemic Mt. Tamalpais manzanita (*Arctostaphylos montana*), leather oak, and Sargent cypress occur on serpentine, while Eastwood's manzanita (*A. glandulosa*), hoary manzanita, chamise, interior live oak scrub, Douglas fir–tanoak, California bay, coast live

oak, and myriad other vegetation types clothe the slopes and valleys off serpentine.

Eastside Lands

Open woodlands of blue oak and foothill pine mingle in the foothill slopes and valley bottoms. Grasslands have wild oats, ripgut brome (*Bromus diandrus*), soft brome, California poppy (*Eschscholzia californica*), Menzies' fiddleneck (*Amsinckia menziesii*), and other early spring wildflowers. California black oak with California fescue in the understory grows at higher elevations. Brewer oak (*Quercus garryana* var. *fruticosa*) forms large patches at somewhat higher elevations. To the north, forests of California bay on north-facing slopes contrast with those of foothill pine forests on south-facing slopes. California buckeye and interior live oak cover steep, rocky slopes. With elevation, Douglas fir–ponderosa pine forests become common. Extensive, pure knobcone pine stands exist scattered on upper slopes and ridges. Incense cedar, sugar pine, and white fir appear in Douglas fir–ponderosa pine forest stands in higher elevations.

Serpentine outcrops with deep, more-productive soils support the same forest types. Sites with thin, less-productive soils support chaparral similar to that farther south. These are the lands of Bear Valley—famous for its spring wildflower shows—and the Mendocino National Forest's Frenzel Creek

Top: Brewer oak, known for its prostrate growth form, grows commonly in Mendocino National Forest.
Julie M. Evens

Middle: Open blue oak woodland in the Cache Creek Wilderness east of Clear Lake.
Robert Wick

Left: An early spring wildflower display at Lacks Creeks, Humboldt County.
Jennifer Kalt

Top: Diverse wildflower fields of Bear Valley in the inner North Coast Range, one ridgeline west of the Great Central Valley.
Ron Maertz

Above: Rare Fritillaria pluriflora *in Bear Valley.*
Roxanne Bittman

Left: Fire-dependent knobcone pine typically grows in an even age stand, reflecting a stand-replacing fire regime common in many regions of the North Coast.
Jim Xerogeanes

Top: Clear-cut logging practices have historically been common along the North Coast, here along the Mad River in Humboldt County.
Jennifer Kalt

Above: Columbian black-tailed mule deer (Odocoileus hemionus columbianus) *doe and fawn north of the Yolla Bolly Wilderness in Mendocino County.*
Mary K. Hanson

Research Natural Area, with choice groves of both McNab and Sargent cypresses.

Logging and Ranching

Logging, more than any other activity, defines much of the North Coast. Today most of the lands in the redwood region are industrial timberlands. The old-growth redwood forests are all but gone, as is mature second-growth established before World War II. Trees in today's timberlands are mainly those of young, third-growth forests.

Preservation of redwood forests began with the establishment of the Muir Woods National Monument in 1908, followed by the Del Norte, Humboldt, Jedediah Smith, and Prairie Creek Redwoods state parks in 1928. Redwood National Park, established in 1968, was expanded in 1978. Today the park and its surrounding state parks, including the recent Mill Creek addition to Del Norte Redwoods State Park, create a grand, uninterrupted redwood wildland of original and maturing second-growth forests.

Changes in the more interior Douglas fir–tanoak forests have been as profound. Before the lumber boom of the mid-1900s, tanoak was the principal tree of commerce. Bark was harvested for its tannic acid, which is used in tanning hides. Douglas fir had its local uses, but logging in earnest required a good transportation network and a demand. These two influences came together with the lumber boom after World War II. Ranchers and small landowners, often called nonindustrial timber owners, "high graded" (cut all trees of value) the large, old Douglas fir trees and left mainly hardwoods, especially madrone and tanoak, creating today's mixed evergreen forests where Douglas fir is less common.

The federal lands of the Mendocino and Six Rivers national forests and those of the BLM have a different history. The agency mandate is a complex balance of supplying timber and protecting forest resources. The need to supply timber increased greatly after World War II. Active road building and vigorous clear-cutting followed, especially at middle elevations. By the 1990s the federal lands became a maze of roads through a checkerboard of original forests, shrublands, and young plantations. Many people predicted that soon all the original forests would be gone, along with their biodiversity.

These concerns resulted in the Northwest Forest Plan of 1998 that substantially reduced the annual cut and changed forestry practices. Today the landscape is healing.

The second mandate, to preserve wild landscapes free of roads and other trappings of civilization—areas "untrammeled by man"—has led to the creation of eleven wilderness areas on the North Coast. The first, the Yolla Bolly–Middle Eel Wilderness, came with the original Wilderness Act of 1964. The California Wilderness Act of 1984 expanded the Yolla Bolly–Middle Eel Wilderness and created North Fork and Snow Mountain wilderness areas. The Northern California Coastal Wild Heritage Wilderness Act of 2006 expanded Snow Mountain and Yolla Bolly–Middle Eel wilderness areas and created eight more on both U.S. Forest Service and BLM lands: Cache Creek, Cedar Roughs, King Range, Mount Lassics, Rocks and Islands, Sanhedrin, South Fork Eel River (two separate units), and Yuki wilderness areas. Maintaining wildlands is the best way to preserve the region's biodiversity.

Livestock grazing since the 1850s has also been important in shaping the region's landscapes, and ranching continues to be an important part of the region's economy. Large ranches, centered on the prairies and oak woodlands east of the redwoods, inhabit the area's ridges and southern slopes where soils are fine-textured and conducive to grass growth. These prairies and woodlands make up the Bald Hills in the Redwood Creek, Mad, Eel, and Mattole river watersheds. Here early ranchers found pastures of perennial grasses for their cattle and sheep, but today introduced annuals dominate. The associated oak woodlands have a history very similar to that of the adjacent prairies. Early ranching in the late 1800s was detrimental as it replaced the frequent, low-intensity fire regime practiced by Native Americans, favoring the ascendancy of Old World annuals. Today, grazing intensity is more sustainable on many of the ranches, while other properties go ungrazed by livestock. This history has encouraged a Douglas fir invasion into the prairies and woodlands in the Bald Hills.

Conservation in the North Coast

Poorly planned development in the rural parts of the North Coast is a major threat to the region's wildlands. Unlike other parts of the state, the North Coast still has extensive landscapes absent of towns and other developments. These

Top: The Bald Hills have a recent history of grazing—and a longer history of low-intensity fires that maintained the prairies for wildlife herds and high plant diversity.
John O. Sawyer

*Above: Female northern spotted owl (*Strix occidentalis caurina*) near Korbel, Humboldt County.*
Mark Herse

Above: The view north from the Cedar Roughs trail, looking out at some of the area recently designated and protected within the Berryessa Snow Mountain National Monument.
JIM ROSE/TULEYOME.ORG

Opposite: The view from North Yolla Bolly Peak looking south over the entire Yolla Bolly–Middle Eel Wilderness.
MICHAEL E. KAUFFMANN

are areas where only dirt roads and trails are available, and while ranchlands are seen, vineyards and rural subdivisions are mostly absent. The maintenance and expansion of these wildlands and the creation of others offers us the best chance to preserve more of the region's wild heritage.

Successful preservation requires thoughtful management, and in most cases means cooperation since large land areas involve several, even many, owners. Even the grand Yolla Bolly–Middle Eel River Wilderness spans the lands of both the U.S. Forest Service and the BLM, as do other new wilderness areas in the Eel. In addition, these areas include private ranchlands.

Save the Redwoods League has started an ambitious conservation project—a corridor from the redwoods to the sea—which will link the wildlands of Humboldt Redwoods State Park with those of the King Range National Conservation Area. They are also expanding BLM's Gilham Butte Late Seral Reserve through land purchases and easements with the cooperation of local owners in the upper reaches of the Mattole River, in an effort to create a wild landscape the partners will manage for maximizing the area's natural values. Upon proposal by the Sierra Club and many others, Berryessa Snow Mountain National Monument, established in 2015, connects the lands from Berryessa Peak to those of the Snow Mountain Wilderness. Partners include federal, state, and non-governmental agencies along with private ranch owners in the area. Successful efforts, like these and others, continue to preserve the natural heritage of the North Coast.

JOHN O. SAWYER

CHAPTER THREE

Klamath Mountains

The Klamath Region contains one of the most exceptional temperate coniferous forests in the world, supporting 100 tree species including 39 conifers and 22 oaks, among a vascular flora of over 3,000 kinds of plants. This species richness is possible because of the region's diverse climates, topography, and geology; a flora with long-enduring lineages; and a biogeography that touches the Cascades, Sierra Nevada, and Coastal Ranges.

The region's climate generally is Mediterranean with dry, hot summers and cool, wet winters. Snow is occasional and fleeting at the lowest elevations. Its many mountain ranges hold deep snow packs well into the summer, supplying much of the water to the region's three major watersheds. The Smith River watershed lies west of the Siskiyou Mountains in northwestern California and is the rainiest in the state (up to 3 m annually). The canyons of the main stem of Klamath River and its tributary, the Salmon River, are sandwiched between the Siskiyou Mountains and the Marble and Salmon mountains to the east. The Scott River watershed, the easternmost tributary of the Klamath River, has the most continental and arid climate in the region. It lies inland from the ocean in the rain shadow of the Marble, Salmon, and Siskiyou mountains. Here summers are warm and winter temperatures are often well below freezing with long-lasting snow. South of the Salmon Mountains and Trinity Alps lies the well-watered Trinity River watershed, the largest tributary of the Klamath River. The Sacramento River watershed is the easternmost in the region. It lies in the rain shadow of the Trinity Mountains, but as winter storms move up the Sacramento Valley, it is also well-watered.

The broad vegetation pattern is similar to that in the Sierra Nevada: a set of elevation belts of foothill, montane, subalpine, and alpine vegetation. However, foothill woodlands and low-elevation chaparral are extensive only in the eastern Klamath, particularly in the southern foothills of the Trinity Mountains.

Top: In the headwaters of Russian Creek in the Russian Wilderness, subalpine fir (Abies lasiocarpa *var.* lasiocarpa)*, Shasta red fir, mountain hemlock, and western white pine decorate the area.*

Above: Emerging from a serpentine mountaintop in Siskiyou Wilderness is Siskiyou fritillaria (Fritillaria glauca)*.*
MICHAEL E. KAUFFMANN PHOTOS

Opposite: Subalpine region on Mount Eddy.
ROBERT WICK

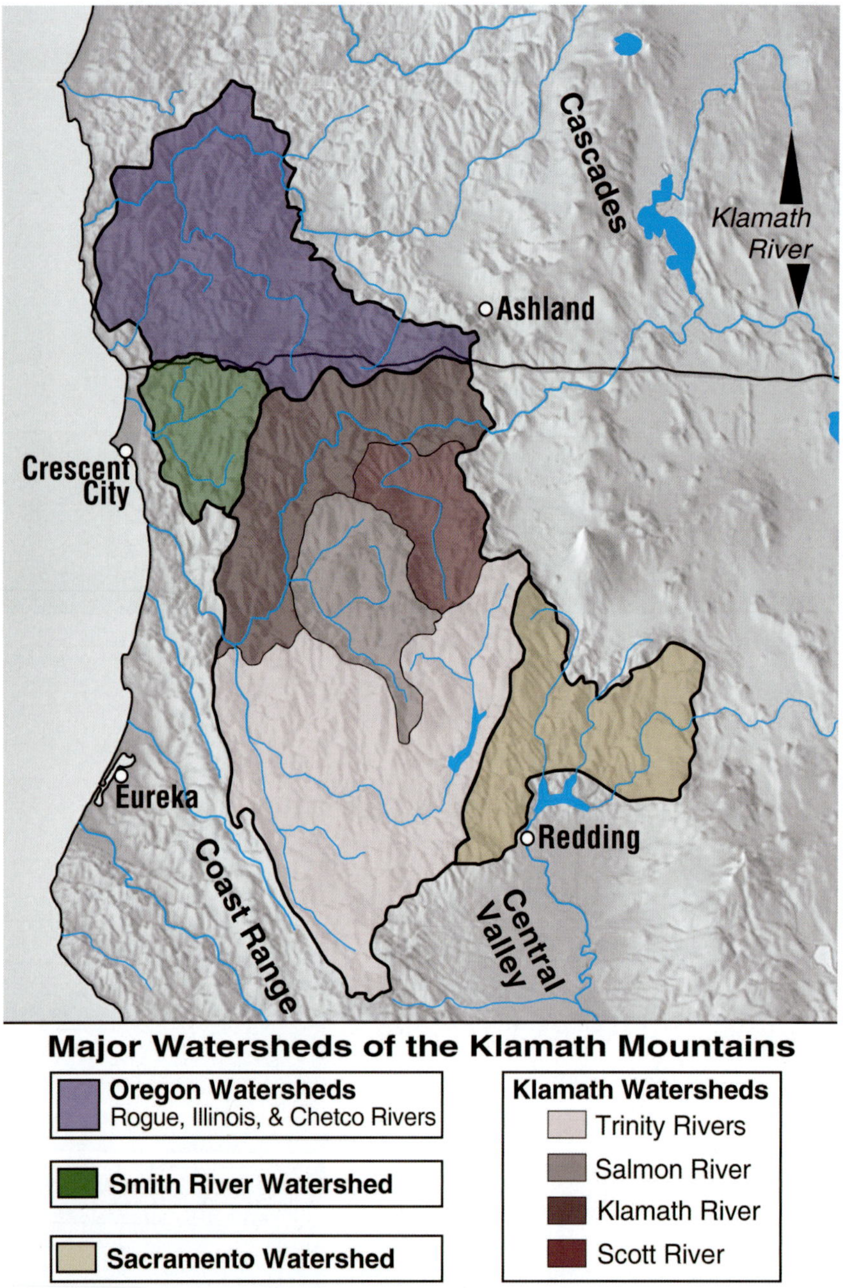

Elsewhere this belt is dominated by forests; trees and other species considered "montane" in other parts of the state grow here at the lowest elevations. The main tree species of the higher-elevation montane and subalpine forests are similar to nearby regions, but in the Klamath they often exist as isolated patches rather than as a continuous forest. Even so, trees grow to the tops of the highest peaks and herbaceous plants that grow elsewhere in the alpine occur here at subalpine elevations.

The Broad Forest Pattern

We might best understand the region's diverse vegetation patterns by first considering those forests with closed canopies that occur on well-developed soils, and whose dominant tree species have high colonizing abilities, long life, and wide ecological tolerances to environmental conditions. Only seven of the region's 100 tree species have these qualities. Douglas fir (*Pseudotsuga menziesii*) and tanoak (*Notholithocarpus densiflorus*) dominate the maritime-influenced forests in the western portions of the region. The more continental eastern part of the region has forests of Douglas fir and ponderosa pine (*Pinus ponderosa*), or foothill pine (*Pinus sabiniana*). Extensive forests dominated by white fir (*Abies concolor*) distinguish montane forests above the low-elevation belt. The notable presence of white fir indicates the low-elevation/montane boundary while the higher montane/subalpine boundary is characterized by either noble fir (*Abies procera*) or Shasta red fir (*A. magnifica* var. *shastensis*). Forests with mountain

Top: Map of the major watersheds of the Klamath Mountains.
John O. Sawyer and Michael E. Kauffmann

Left: Foothill pine growing in the eastern portion of the region.
John O. Sawyer

Above: Patchy mixed fir forest grows up to the peaks of the Marble Mountains.
Jeff Bisbee

Left: Noble fir grows at higher elevations of the montane/subalpine boundary.
John O. Sawyer

Montane chaparral with tobacco brush gives over to mountain hemlock stands on the north-facing slopes along the trail to East Weaver Lake in the Trinity Alps Wilderness, Trinity County. John O. Sawyer

hemlock (*Tsuga mertensiana*), noble fir, Shasta red fir, or diverse woodlands with various pines are indicative of the subalpine belt.

Sparse woodlands of stunted trees and barrens, even in the wettest parts of the region, interrupt this otherwise broad forest pattern. This vegetation occurs on ultramafic rocks (peridotite, serpentinite, and associated rock types). Peridotite is a coarse-grained rock high in magnesium and iron. When exposed to air, the rock surfaces slowly become brick red as the iron oxidizes. When groundwater interacts with the minerals in peridotite, they undergo a special process of hydration that creates the serpentine minerals found in serpentinite. These ultramafic rocks, which botanists refer to as "serpentine," occur throughout northwest California, the Central Coast, the Sierra Nevada, and other parts of the world. The Klamath Region has the largest serpentine outcrops in North America. These outcrops exist over a wide range of elevations and climatic conditions, and they contribute significantly to the region's uniqueness.

Soils that develop from ultramafic rocks vary greatly in soil depth and water-holding capacity, but all are predictably sterile. For many plant species the soils contain high, even toxic, levels of heavy metals (boron, chromium, nickel). On the other hand, many plants have evolved to grow under these harsh conditions, and some are restricted to them.

The overall appearance of plants growing on serpentine is distinctive. Woody plants have small, stiff, long-lasting evergreen (sclerophyllous) leaves. These plants are short and have extensive root

systems, suggesting that most of their energy goes into maintenance and herbivore defense, leaving only minimal energy left for slow growth. Herbaceous plants are annuals or slow-growing perennials.

Soils that develop from other rock types in the region are more productive. Granite soils are usually coarse, well-drained, and low in nutrients, but have enough macronutrients for plant growth. Given enough precipitation, granite soils can support forest trees and deep-rooted shrubs. The region has many other nutrient-rich substrates like igneous, sedimentary, and metamorphic rock types (including limestone). All support forests similar to those on granitic soils.

Top: Jeffrey pines dapple the stark landscape on serpentine ridgelines in the mesic Siskiyou Wilderness.

Above: Mixed conifer fire mosaic in the southern Siskiyou Wilderness.
Michael E. Kauffmann photos

Fire History

Lightning is common in the Klamath Mountains. Often it is associated with

*Top left: Post-fire chaparral with tobacco brush and various herbs including fireweed (*Chamerion angustifolium *ssp.* circumvagum*) and penstemon.* Julie M. Evens

Top right: Stunted whitebark pine on Mount Eddy, looking east toward Mount Shasta. Jeff Bisbee

Above: Fire-adapted knobcone pines are common across the Klamath Mountains. Michael E. Kauffmann

summer rainstorms, but it is the episodes of "dry lighting" that can ignite hundreds of fires. These fires expand, unite, and may burn for months in large areas until extinguished later in the season by rain and snow. For example, the Megram Fire in the western Trinity Alps in 1999 burned more than 57,000 ha, while the many fires in the Trinity Alps and Trinity Mountains in 2008 totaled 118,250 ha. These fires tend to be moderately to highly complex surface fires. They generally burned through duff and litter on the forest floor while occasionally reaching the canopy, killing patches or whole stands of trees—surprisingly, many trees survived. In contrast, areas of chaparral burned intensively, devouring all shrubs in the stand.

Ecologists believe most stands in the region have experienced at least several fires each century. The general forest fire regime was one of frequent, low- to moderate-intensity fires. Such a regime creates a complicated mosaic of chaparral, forest, and woodland stands of various ages and sizes throughout the region. This overlay on the region's diverse climates, topography, and geology obscures the broad forest pattern. The vegetation patterns can be better understood by subdividing the area into watersheds.

The Sacramento River Watershed

The headwaters of the Sacramento River form in the Trinity Mountains at Mount Eddy, at 2,750 m the highest peak in the region, and on the lower slopes of Mount Shasta in the adjacent Cascades. From subalpine lakes and mountain meadows, stream channels surge through rugged, rocky canyons to Shasta Lake, the largest reservoir in the state, before entering the Sacramento Valley and ultimately to San Francisco Bay. The rounded ridges and mountain slopes are covered with dark forests and well-lighted woodlands or chaparral. Extensive limestone areas offer high species richness, interesting fossils, and secret caves. The granitic Castle Crags rise above the forest mantle, and ultramafic substrates offer their own special plants.

Stands of the ponderosa pine–Douglas fir forest alliance make up the most extensive forest type at lower elevations in the watershed. Bigleaf maple

Top left: Foothill pine mixing with chaparral above Whiskeytown Lake, Shasta County.
Jeff Bisbee

Above: Ponderosa pine–incense cedar forest alliance creates mixed stands at mid-elevations.
John O. Sawyer

Left: A California pitcher plant fen in the Sacramento Watershed.
Michael E. Kauffmann

(*Acer macrophyllum*) creates sizable stands on steep, north-facing slopes. Extensive stands of knobcone pine (*Pinus attenuata*) or foothill pine form a mosaic with chaparral alliances containing chamise (*Adenostoma fasciculatum*), deer brush (*Ceanothus integerrimus*), or white-leaf manzanita (*Arctostaphylos viscida*). Mixed stands of interior live oak (*Quercus wislizeni*) and madrone (*Arbutus menziesii*) mix with those of canyon live oak (*Quercus chrysolepis*). Valley oak (*Quercus lobata*) reaches its northern limit here.

The white fir forest alliance forms pure forests. Mixed stands with ponderosa pine–incense cedar (*Calocedrus decurrens*), or Douglas fir and ponderosa pine, each separate alliances, are found at mid-elevations. On drier sites, the California black oak (*Quercus kelloggii*) forest alliance occurs, often with scattered trees of ponderosa pine. Patches of bitter cherry (*Prunus emarginata*) or green leaf manzanita (*Arctostaphylos patula*) are also common. At higher elevations, Shasta red fir first enters and then becomes the only forest tree thus defining the red fir (*Abies magnifica*)

Port Orford cedars survive in disjunct stands, here in the Cedar Lake Basin in the headwaters of the Sacramento River.
Michael E. Kauffmann

forest alliance. Shallow serpentine soils support the Jeffrey pine (*Pinus jeffreyi*) forest alliance among swaths of wedge leaf ceanothus (*Ceanothus cuneatus*) at low elevations. The highest ridges and peaks mostly support woodlands of foxtail pine (*Pinus balfouriana*), Jeffrey pine, and whitebark pine (*Pinus albicaulis*) on gabbro and ultramafic substrates. The more mesic Port Orford cedar (*Chamaecyparis lawsoniana*) forest alliance often contains lodgepole pine (*Pinus contorta* ssp. *murrayana*), mountain hemlock, Shasta red fir, western white pine (*Pinus monticola*), and pink mountain-heath (*Phyllodoce empetriformis*). Mountain alder (*Alnus incana*) often lines the higher elevation creeks.

The Klamath River Watershed

From its headwaters in the Cascades of southern Oregon, the Klamath River also has a long and circuitous route to the Pacific Ocean. The river enters the Klamath Region just west of Interstate 5 and from there travels west and south until it cuts through the Siskiyou Mountains and takes a hard turn to the north towards the Pacific Ocean. Here the river flows through alternating scenes of gorges, steep forested slopes, and small river terraces cutting through the Siskiyou Mountains to the west and Scott Bar and Marble Mountains to the east. The higher slopes support deep and dark fir forests sheltering extensive meadowlands and glacial lakes.

The Scott River tributary flows through the largest valley in the region. Scott Valley seems like a taste of the Great Basin with its extensive ranchlands. Woodland alliances of Oregon white oak (*Quercus garryana*) and western juniper (*Juniperus occidentalis*) are widespread. Both woodlands contain big sagebrush (*Artemisia tridentata*) and rubber rabbitbrush (*Ericameria nauseosa*). Water birch (*Betula occidentalis*) grows along the

Above: The Russian Wilderness, home to the Miracle Mile, holds 18 conifer species in one square mile. This is one of the most species-rich temperate coniferous forests on Earth.
Michael E. Kauffmann

Right: Scott Valley.
Jennifer Whipple

Above: White corn lily meadows commonly occur in the region, here in the Marble Mountains. Jeff Bisbee

watercourses, as it does throughout the interior West.

Looming to the west of the Scott Valley, the Marble and Salmon mountains offer a landscape similar to that of the Trinity Mountains. Forest stands at montane elevations are mainly those of white fir forest and ponderosa pine–incense cedar forest alliances. Patches of the green leaf manzanita chaparral and huckleberry oak (*Quercus vacciniifolia*) chaparral alliances also contain tobacco brush (*Ceanothus velutinus*), oceanspray (*Holodiscus discolor*), and shrub tanoak (*Notholithocarpus densiflorus* var. *echinoides*). Shasta red fir becomes common with elevation, sheltering lush meadows of meadow barley (*Hordeum brachyantherum*) patches, white corn lily (*Veratrum californicum*) patches, and herb-rich (*Senecio triangularis*) meadow alliances. The highest ridges and peaks support stands of Shasta red fir and mountain hemlock, with local areas of foxtail pine, Jeffrey pine, lodgepole pine, western white pine, and whitebark pine.

The Salmon River tributary flows through the most rugged and isolated watershed in the region. Its two branches flow though steep-walled canyons where canyon live oak mingles conspicuously with the ponderosa pine–Douglas fir forest and California black oak forest upstream. Forest stands of Douglas fir and tanoak alliance are common downstream. Canyon live oak becomes abundant on the driest, rocky sites, and Oregon white oak inhabits the thinnest of soils. Forest stands at montane elevations are commonly pure Douglas fir, white fir, or mixtures of these conifers that represent the white fir forest, white fir–Douglas fir forest, and Douglas fir forest alliances. Patches of

Sadler oak (*Quercus sadleriana*) replace those of huckleberry oak as the common montane chaparral alliance. With increased elevation, Brewer spruce (*Picea breweriana*) forests are found as isolated pockets among red fir forests and mountain hemlock forests—which become common and extensive at subalpine elevations. The surrounding terrain is one of crags, tarns, creeks, and associated meadows of the mountain alder thickets, showy sedge (*Carex spectabilis*) sod, white mountain heather (*Cassiope mertensiana*), Parry's rush (*Juncus parryi*) outcrops, mountain sorrel (*Oxyria digyna*) patches, heretic penstemon (*Penstemon heterodoxus*) patches, mountain pride (*P. newberryi*) patches, patches of Tolmie's alpine saxifrage (*Saxifraga tolmiei*), and dwarf bilberry (*Vaccinium caespitosum*) alliances.

The Trinity River tributary forms the grandest watershed in the region. Headwaters flow from the western slopes of Mount Eddy and then south through the Trinity Alps to meet its South Fork near Willow Creek. The combined waters flow north to meet the Klamath River at Weitchpec. Upstream from the mouth of the South Fork, the scene is similar to that surrounding the Sacramento River where stands of the foothill pine woodland, Douglas fir forest, and ponderosa pine–Douglas fir forest alliances clothe the slopes. Canyon live oak often line the canyon walls above riparian vegetation including white alder (*Alnus rhombifolia*), black cottonwood (*Populus trichocarpa*), and sandbar willow (*Salix exigua*) at lower elevations. Mountain alder can be found in the higher elevation riparian areas.

In recently burned areas, swatches of the wedge leaf ceanothus chaparral and deer brush (*Ceanothus integerrimus*) chaparral alliances pioneer the recently

The high ridges and peaks support woodlands of foxtail pine on gabbro and ultramafic substrates, here in eastern Trinity Alps Wilderness.
Michael E. Kauffmann

cleared landscape. At montane elevations, Douglas fir, ponderosa pine, and white fir mix with incense cedar. On drier sites, California black oak associates with conifers. Patches of huckleberry oak chaparral, Brewer oak (*Quercus garryana* var. *fruticosa*) scrub, green leaf manzanita chaparral, or common whiteleaf manzanita (*Arctostaphylos manzanita*) are common as well as a high diversity of dry and wet meadow types.

Forest stands of the Douglas fir–tanoak alliance are common at lower

Top: Jeffrey pine trees frame Condrey Mountain along the Pacific Crest Trail in the Scott Mountains of Shasta-Trinity National Forest.
John O. Sawyer

Left: Mountain alder near Twin Lakes in the Russian Wilderness.
Julie M. Evens

Above: The pendulous nature of Brewer spruce starkly contrasts with the surrounding conifers and Thompson Peak Glacier in the background at Grizzly Lake in the Trinity Alps Wilderness.
JEFF BISBEE

Left: Beaked sedge (Carex utriculata) *and blister sedge* (Carex vesicaria) *with yellow pond-lily* (Nuphar lutea ssp. polysepala) *mats in a Trinity Alps Wilderness aquatic habitat.*
ROBERT WICK

Above: Riparian vegetation along the Trinity River includes white alder and black cottonwood. Robert Wick

Right: A dry meadow in the Trinity Alps includes scarlet gilia (Ipomopsis aggregata) *and arrowleaf buckwheat* (Eriogonum compositum) *with foxtail pine in the background.* Julie M. Evens

Opposite: Marsh marigold (Caltha leptosepala), *alpine shootingstar* (Dodecatheon alpinum), *and willow bloom in a wet meadow in the Trinity Alps.* Robert Wick

elevations, downstream from the South Fork's mouth. Port Orford cedar makes its western occurrence in these forests. Stands of the canyon live oak forest, Oregon white oak woodland, and California black oak forest alliances often contain Douglas fir on the drier slopes. At montane elevations, stands of the white fir–Douglas fir forest, white fir forest, and red fir–white fir forest alliances produce shady forests. As elevation increases in the central granitic "White" Trinity Alps, mountain hemlock enters, becoming a common member of the red fir forest and mountain hemlock forest alliances at subalpine elevations. The eastern "Red" Trinity Alps, with a foundation of gabbro and serpentine, are a treasure-trove of rare plants and rich woodlands of foxtail pine, lodgepole pine, Shasta red fir, and western white pine. In the west, lower

Above: Mountain hemlock in the White Trinity Alps.
Robert Wick

Right: Stands of the white fir–sugar pine forest alliance along the trail to Union Lake in the Trinity Alps Wilderness.

Opposite top: Old Jeffrey pine trees grow on peridotite at High Divide in Six River National Forest.

Opposite bottom right: Huckleberry oak chaparral covers the granite below stands of Shasta fir near East Weaver Lake in the Trinity Alps Wilderness.
John O. Sawyer photos

elevation Green Alps drained by the New River are mostly forested with Douglas fir, noble fir, and white fir.

The South Fork tributary is bounded by the Trinity Mountains on the east, the Yolla Bolly Mountains on the south, and South Fork Mountain on the west. The land is lower than the Trinity Alps, but equally rugged. Forest stands of the Douglas fir–tanoak alliance change to the typical eastern mixture of the ponderosa pine–Douglas fir forest alliance. Jeffrey pine and its own set of serpentine plants and seeps characterize the ultramafic substrates. Blue oak (*Quercus douglasii*) grows in the Hayfork and Hyampom valleys.

The Smith River Watershed

The free-flowing Smith River, with its 300 miles (483 km) of famous green

water, is the crown jewel of the National Wild and Scenic River System, promoting world class fishing and whitewater sports. The watershed has a dual nature. On one side, extensive ultramafic substrates of the North Fork support open Jeffrey pine forests; dense shrublands of shrub tanoak chaparral, Brewer oak scrub, and huckleberry oak chaparral; barrens of Idaho fescue (*Festuca idahoensis*) grassland; and soggy California pitcher plant (*Darlingtonia californica*) fens. Species like the federally-listed McDonald's rock cress (*Arabis macdonaldiana*) contribute to the region having the highest number of endemic vascular plants of any serpentine landscape in North America.

The South Fork of the Smith River, where serpentine is less common, boasts dense stands of Douglas fir–tanoak forest merging into those of the white fir–Douglas fir, white fir, and red fir–

white fir forest alliances with elevation. Knobcone pine also occurs in patches with chaparral, typically in areas with recent fire disturbance. Port Orford cedar forest forms great stands in the upper valleys beside species-rich meadows and fens of the western false asphodel–California bog asphodel (*Triantha occidentalis–Narthecium californicum*) fens alliance. Brewer spruce is joined by Alaska yellow-cedar (*Callitropsis nootkatensis*) and mountain hemlock in the high Siskiyou Mountains. Sitka alder (*Alnus viridis*) lines the higher streamsides.

Vegetation Persistence and Conservation

In many areas, the expansive landscapes of the Klamath Mountains are not highly fragmented with roads. Only a small number of highways follow the rivers, and few Forest Service roads permit access to the more moderate terrain at middle elevations, where logging was most intense in the mid-1900s. These logging roads lead to trailheads that provide access to the region's eight wilderness areas.

Wilderness promotes diversity by isolating species from invasion events.

Above: Alaska yellow-cedar mixes with white fir and Shasta red fir in the higher elevations of the Siskiyou Mountains. Jeff Bisbee

Opposite top: Knobcone pine forest with an understory of chaparral in the Smith River National Recreation Area.

Opposite bottom: Decorated in winter snow, Preston Peak in the Siskiyou Wilderness is within the Smith River watershed. Michael E. Kauffmann photos

Roads serve as vectors for disease and non-native plants—thus increasing a need for management by humans. Management comes at a cost both financially and to the systematic interactions of the organisms in an ecosystem. Let the wild be wild; the best way to preserve regional biological diversity is to set aside large land areas unfragmented by roads.

While these wilderness areas tend to protect the less productive, high-elevation terrains, the roadless areas outside the existing wilderness often support more productive habitats and forests. These lands, considered together, form de facto wilderness areas twice the size of the existing protected wilderness. For example, the lands of the Trinity Alps, Russian, Marble, Red Butte, and Siskiyou wildernesses and adjacent areas function today as one great wild landscape, regardless of designation. The Klamath Region, one of the largest intact forested areas in the nation, deserves a strong conservation strategy to protect its rich floristic and vegetation diversity.

Above: Bear Lake, in the Trinity Alps Wilderness.
Jeff Bisbee

Left: John Sawyer below Emerald and Sapphire Lakes, Trinity Alps Wilderness.
Julie M. Evens

Opposite: Douglas fir forest lines slopes of the Klamath River above Somes Bar.
John O. Sawyer

John O. Sawyer
Michael E. Kauffmann

THE
AERMOTOR
CHICAGO

CHAPTER FOUR

Modoc Plateau and Northwestern Basin and Range

In California's remote northeastern corner lies a vast volcanic plain of open sweeping vistas known as the Modoc Plateau and Northwestern Basin and Range. Extensive basalt flows 200 m deep, volcanic craters and obsidian domes, ash and pumice deposits, lava tubes, basalt pools, talus jumbles, and lava rims, provide evidence that volcanic activity in the region has been massive and continuous for the last 60 million years. Three active volcanos—Medicine Lake, Mt. Shasta and Lassen Peak—loom over the region's western edge. The gradual rise of the Warner Mountains, one of several north-south trending mountain ranges thrust upward along tectonic faults, defines the region's eastern margin (Norris and Webb 1976).

The Pit River flows gradually west and south through this broad volcanic landscape, draining 60% of the region (1.2 million ha), and contributing over 20% of the total volume of water into the Sacramento River, California's largest river. Extensive basalt flows act as a giant sponge, absorbing massive amounts of water from melting snow and rainfall. This water percolates down through the porous volcanic rock until it reaches more impenetrable underlying layers, causing some of this water

Above: The Warner Mountains and Surprise Valley form the northeastern boundary of the region.

Opposite: Windmill in sagebrush scrub on the Madeline Plains.
MIKE DOLAN PHOTOS

Above: Basalt rim above the Pit River, east of Fall River Mills, including fall color of California black oak and cherry on slopes and willow along the river.
Robert Wick

Below: Sandhill crane at Modoc National Wildlife Refuge.
Todd Keeler-Wolf

to mysteriously disappear underground and creating a tangle of lost rivers and lava tubes across the region. These subterranean waters slowly flow southward, finally emanating in voluminous springs of crystal clear water in such places as Ahjumawi and Big Valley springs. These springs pour into the Pit River, filling it with water that has traveled through underground aquifers for a hundred years or more.

Huge inland lakes scattered throughout the region are remnants of a vast inland seaway that once connected the Modoc Plateau to the Pacific Ocean. Pleistocene lakes—such as Honey Lake, Tule Lake, and Goose Lake—shrink and swell with fluctuating rainfall, sometimes drying out completely. Smaller ephemeral lakes and vernal pools, filled by snow melt rather than rainfall, occur in low lying flats and shallow basins, providing brilliant multi-colored displays of vernal pool flowers every spring. These vast wetlands provide numerous breeding grounds for spectacular flocks of water birds such as western and eared grebes (*Aechmophorus occidentalis, Podiceps nigricollis*), sandhill cranes (*Grus canadensis*), Caspian terns (*Hydroprogne caspia*), and American white pelicans (*Pelecanus erythrorhynchos*).

Traveling into the Modoc Plateau and Northwestern Basin and Range from other parts of California, familiar vegetation slowly becomes interspersed with unusual species. Many Californian species reach the northeastern extent of their ranges here, including California black oak (*Quercus kelloggii*) and vernal pool species typically associated with California's Central Valley. Heading north and east, we are increasingly surrounded by low sagebrush (*Artemisia arbuscula*) and perennial grasslands more typical of the Great Basin. At higher

elevations, we begin to see western juniper (*Juniperus occidentalis*) and pine (*Pinus jeffreyi, P. ponderosa*) woodlands, vegetation similar to the high deserts of eastern Oregon and Nevada. We are entering the western extent of many of these species ranges, a transition zone between floras with unique combinations of species found nowhere else. Spending time in this region takes us back in time to an era when the landscape was shaped mostly by natural forces. Urban development accounts for less than one percent of the land area here; undeveloped lands stretch in every direction as far as the eye can see. This is the least populated part of Northern California, with less than 10,000 people in all of Modoc County. A single stop light in the town of Alturas gives us pause to appreciate the isolated and remote nature of this region.

Although modern inhabitation is low, this was one of the longest continually

Top: Goose Lake, on the Oregon–California border, is a closed basin lake with no outflow. Water levels fluctuate with precipitation and evaporation leading to high alkalinity.
Kyle Merriam

Above: Perennial grassland species (such as Leymus cinereus*) in the Great Basin are common in the region.*
Todd Keeler-Wolf

occupied areas in North America. The history and cultural legacy of Native American people in the region stretches back thousands of years, left behind in countless rock art and archeological sites. Close inspection of the ground reveals arrowheads and obsidian flakes scattered everywhere. Using the extensive lava tubes and caves as refuge, the Modoc Indians waged the only major Indian War fought in California. After winning several battles, the war was ultimately lost by the Modoc in the late 1800s. However, evidence of their life here persists. The battlefields of this conflict, known as the Modoc War, are largely located within Lava Beds National Monument and are still preserved today.

Climate

The region's climate, classified as Mediterranean Montane (Bailey 1995), is colder and drier than most other areas of California. Average annual precipitation ranges between 28 and 100 cm, falling primarily as rain or snow between October and May, followed by a dry summer season. Precipitation comes primarily from winter Pacific storms, although the rain shadow effect of the Southern Cascade Mountains to the west blocks much of this precipitation from actually reaching the region. Average annual temperatures are mild, with winter lows generally below freezing, and summer maximum temperatures rarely exceeding 33° C (WRCC 2012). Lower elevations generally experience warmer temperatures with lesser amounts of rainfall; higher elevations experience

Top: Belfast Petroglyphs on Willow Creek near Susanville.
Robert Wick

Left: Ponderosa pine woodland after a summer hailstorm not far from the Butte Creek crossing of Highway 44.
Don Lepley

cooler temperatures and greater amounts of precipitation and snowfall. The region can experience isolated summer storms, and although these monsoons are rare, they can have significant localized effects on vegetation. Temperatures here are too low for plant growth during the winter when most precipitation falls, and drought and high temperatures limit plant growth in summer. These climatic patterns cause plant productivity to be generally low throughout the region (Smith and Davidson 2003). The combination of summer drought and periodic lightning storms predisposes much of the area to lightning-caused fires, and many vegetation types found in the region depend on specific fire regimes to persist.

Geographic Extent

The Modoc Plateau and Northwestern Basin and Range span almost two million hectares, extending north to the Oregon border, east to the Nevada border, and south to include Honey Lake near Susanville, California. The region's highest point is the summit of Eagle Peak (3,015 m) in the Warner Mountains. A second prominent mountain range, the Big Valley Mountains, stands to the west. Honey Lake lies at the southern boundary of the region at its lowest elevation of just below 1,219 m. In addition to Honey Lake, the region supports a number of contained inland basins supporting large expanses of alkaline flats, including portions of the Klamath Basin and Goose Lake in the north, the Madeline Plains and Eagle Lake near the center, and the Surprise Valley along the Nevada border east of the Warner Mountains. These basins have no outlets, so freshwater is mostly lost through evaporation, leading to alkaline conditions.

Ownership

Sixty percent of the region (1.2 million ha) is managed by Federal, State, and local agencies, including the Bureau of Land Management (BLM), the U.S. Forest Service (USFS), the U.S. Fish and Wildlife Service (USFWS), and the California Department of Fish and Wildlife (CDFW). Almost 30% of the

Top: Western juniper on snow covered mountains in the South Warner Wilderness.
Michael E. Kauffmann

Above: Summer storms on the Modoc Plateau can have localized effects on vegetation, such as rubber rabbitbrush scrub vegetation, which is fire-adapted and typically is enhanced by or colonizes after lightning-caused fire.
Forest Gauna

A view of Honey Lake from Thompson Peak lookout.
Craig Odegard

area is managed by the USFS and 25% by the BLM, while smaller areas are managed as National Wildlife Refuges by the USFWS and State Wildlife Areas by the CDFW. Although approximately 800,000 ha, or 40% of the region, is privately owned, much of it is used for agriculture (151,136 ha) or as pasture and range land. Despite the relatively undeveloped nature of the region, only nine percent of the land is in designated as reserves, such as wilderness areas. This degree of protection is less than in any other region of California except the Great Valley (TNC 2001).

Threats

Although it is one of the least-developed regions of the state, ecosystems here are considered among the most threatened in North America (TNC 2001). Many of the plant communities have been substantially altered or degraded over the past 150 years by a combination of factors, including overgrazing by livestock and feral horses, invasion of non-native species such as cheatgrass (*Bromus tectorum*), expansion of western juniper, climate change, and altered fire regimes (CDFG 2006). These threats have led to the degradation and loss of perennial grasslands (West 1983), aspen (*Populus tremuloides*) (Di Orio et al. 2005), and meadow and riparian areas (USFS 2001, 1991). Aquatic ecosystems and wetlands have been damaged or destroyed as a result of water diversions, erosion from logging roads, grazing

activities, and introductions of non-native fish and invertebrates (CDFG 2006).

Geology

In addition to the dominant volcanic features of the region, several unique soil types support vegetation types here. The dominant soils—mollisols, formed from extensive deposits of weathered basalt (Smith and Davidson 2003, NRCS 1993)—are generally shallow, rocky, and poorly developed as a result of cool temperatures and low precipitation that provide limited nutrients and water holding capacity for vegetation. Vertisols—very shallow clay soils that form over basalt bedrock and are subject to significant shrink and swell cycles during the year—support vernal pools and ephemeral wetlands in the region. Aridisols—occurring on sites where potential evapotranspiration greatly exceeds precipitation—often accumulate growth-limited salts, supporting only specialized vegetation that can tolerate alkaline conditions.

Vegetation Types

The vegetation of the Modoc Plateau and Northwestern Basin and Range is loosely segregated along elevational gradients. At the lowest elevations, pluvial lakebeds form habitat for alkaline scrub species. Above the pluvial basins sagebrush (*Artemisia* sp.) steppe dominates. Western juniper is found in the transition zone between the sagebrush and mixed conifer zones. A wide variety of perennial bunchgrasses occur throughout the region in the understory of shrub and open conifer forest vegetation types. Wetland, spring, meadow, vernal pool, riparian, and aspen communities are scattered across the region and support diverse communities of wildlife. Fire-adapted stands of the serotinous Baker cypress (*Hesperocyparis bakeri*) and knobcone pine (*Pinus attenuata*) can be found on the basalt flows of Timbered Crater.

Top left: Many public and private lands in the region are used for livestock grazing.
MEREDITH GOSEJOHAN

Top right: Cheatgrass invasion has converted thousands of acres of sagebrush steppe to annual grasslands.
PETER WEISBERG

Above: Vertisols in the Devils Garden area of the Modoc Plateau support curly blue grass grasslands.
TODD KEELER-WOLF

Above: Vegetation types are generally segregated along elevational gradients.
Mike Dolan

Opposite top: Eagle Lake supports a variety of alkaline adapted vegetation such as salt heliotrope (Heliotropium curassavicum) *and salt grass* (Distichlis spicata).
Craig Odegard

Opposite bottom: Greasewood and sagebrush near Honey Lake with the Skedaddle Mountains in the background.
Don Lepley

Alkali Desert Scrub

Pluvial basins in the Surprise Valley, Eagle Lake, Goose Lake, Klamath Basin, and Honey Lake areas support a diversity of alkaline-adapted vegetation. These relatively rare vegetation types occupy only three percent of the region, occurring in the lowest elevations below 1,700 m. These inland basins do not have outlets, so fresh water is left to evaporate, leading to an accumulation of salts. The resulting soils are too alkaline for most species of sagebrush, and the dominant shrubs in this community are generally small, grayish, intricately-branched, and often spiny.

Greasewood (*Sarcobatus vermiculatus*) scrub is a common alliance found in alkaline basins including the Honey Lake, Goose Lake, Surprise Valley, Tule Lake, and Klamath Marsh areas. Co-occurring species include fourwing saltbush (*Atriplex canescens*), shadscale (*A. confertifolia*), and winterfat (*Krascheninnikovia lanata*).

Sagebrush Steppe

The sagebrush steppe is a complex of plant communities, where sagebrush co-occurs with diverse assemblages of other shrubs, grasses, and forbs. Sagebrush (*Artemisia* spp.) scrub is the dominant vegetation type throughout the Great Basin, occupying almost 45 million ha. The Modoc Plateau and Northwestern Basin and Range supports the most extensive areas of sagebrush steppe in California—here it occupies over 45% of the land area. This vegetation type occurs on extensive plains in the drier areas, at higher elevations than the alkaline scrub-dominated pluvial basins and at lower elevations than the coniferous forest and oak woodland vegetation types. Sagebrush cover in healthy steppe ecosystems is generally around 10–15%, with drought-adapted native bunchgrasses, such as Idaho fescue (*Festuca idahoensis*) and ashy ryegrass (*Leymus cinereus*), also common. Selective grazing by livestock can promote the cover of sagebrush at the expense of native grasses and forbs (Young et al. 2007).

Low sagebrush (*Artemisia arbuscula*) scrub is common here, and also extends well outside the region. Low sagebrush plant communities occur primarily in the northern Devil's Garden area and the lower slopes of the Warner and Big Valley Mountains. It is dominant at lower elevations of around 1,500 m, but can be found at elevations from 1,229 to 2,879 m. Lahontan low sagebrush (*Artemisia arbuscula* ssp. *longicaulis*), is the most widespread subspecies of low sagebrush in the region, and is associated with shallow, clay textured soils above a restrictive hardpan or bedrock such as that found in the Devils Garden area.

Lahontan low sagebrush is found only in California, Nevada, and Oregon. This subspecies of low sagebrush formerly occupied extensive areas of the Modoc Plateau, but is rapidly disappearing as a result of increasing fire frequency caused by cheatgrass invasion. Other subspecies of low sagebrush are found at higher elevation sites, such as the Warner Mountains. Low sagebrush scrub is one of the most diverse plant communities in the region, including native grasses such as one-spike oat grass (*Danthonia unispicata*), prairie junegrass (*Koeleria macrantha*), bluebunch wheat grass (*Pseudoroegneria spicata*) and Idaho fescue. Low sagebrush scrub also supports numerous species of wildflowers, such as pussytoes (*Antennaria* spp.), fleabane (*Erigeron* spp.), desert parsley (*Lomatium* spp.), balsam root (*Balsamorhiza* spp.), buckwheat (*Eriogonum* spp.), phlox (*Phlox* spp.) and clover (*Trifolium* spp.), creating colorful floral displays in the spring.

Top: Lahontan low sagebrush in the Fitzhugh Creek area. Todd Keeler-Wolf

Above: High elevation low sagebrush in the Warner Mountains. Mike Dolan

Opposite top: Low sagebrush flats support diverse forb and grass communities. Jennie Moore/USFS

Opposite middle: Wyoming big sagebrush at the Likely Tablelands. Mike Dolan

Opposite bottom: Big sagebrush steppe in the Modoc National Forest near Mount Dome in Siskiyou County. Forest Gauna

Big sagebrush (*Artemisia tridentata*) is the most abundant and widespread of all the species of sagebrush, and by some estimates it has more biomass than any other species of shrub in the world (Schultz 2009). The big sagebrush shrubland alliance is very common in the region, but concentrated in the southeast. It is generally found on deep, well-drained soils with well-developed soil horizons at elevations around 1,600 m. Big sagebrush can occur at a range of elevations, from 1,027 to 2,822 m. Common species associated with big sagebrush shrublands include aspen, cherry (*Prunus* spp.), gooseberry (*Ribes* spp.), snowberry (*Symphoricarpos* spp.), bitter brush (*Purshia tridentata*), black sagebrush (*Artemisia nova*), and curl leaf mountain mahogany (*Cercocarpus ledifolius*).

Three subspecies of big sagebrush are dominant here: Wyoming big sagebrush (*Artemisia tridentata* ssp. *wyomingensis*); Great Basin big sagebrush (*Artemisia tridentata* ssp. *tridentata*); and mountain big sagebrush (*Artemisia tridentata* ssp. *vaseyana*). Although these subspecies may hybridize, they generally occupy different habitats. Great Basin big sagebrush, the tallest subspecies, is found on deep, well-drained soils. Settlers in the nineteenth century recognized that the presence of Great Basin big sagebrush indicated good agricultural soils, and as a consequence, most of the habitat for Great Basin big sagebrush was plowed under by the early twentieth century. Mountain big sagebrush is the medium sized, flat-topped shrub found on mountain slopes. Wyoming big sagebrush is the short to medium sized shrub found on harsh, dry sites from lowest elevation in the valleys to mountain slopes. Wyoming big sagebrush is often found in the drier portion of the region and co-occurs with perennial grasses such as Thurber's needlegrass (*Achnatherum thurberianum*) and desert needlegrass (*A. speciosum*). Great Basin big sagebrush communities occupy intermediate elevations and often co-occur with bluebunch wheat grass, Thurber's needlegrass, or ashy ryegrass on productive sites with deep alluvial soils. Mountain big sagebrush communities occupy the most productive sites, usually at higher elevations on deep, well-developed mollisols. These areas generally support higher shrub cover, where mountain big sagebrush co-occurs with bitter brush and Idaho fescue (Young et al. 2007).

One of the most serious threats to the sagebrush steppe is the invasion of cheatgrass and the resulting changes to historic fire regimes caused by cheatgrass.

Over 25,000 ha of the region are mapped as dominated by cheatgrass (USFS 2010). Native to southern Europe, North Africa, and southwestern Asia, cheatgrass was first dispersed in northeastern California sometime in the early 1900s, probably via contaminated grain seeds (Novak and Mack 2001). Cheatgrass displaces native grasses, forbs, and shrubs by more effectively tapping soil moisture than other species due to its early-season growth, and hinders seedling establishment by reducing moisture and nutrients in surface soils (Norton et al. 2004). Overgrazing has reduced the abundance of native perennial grasses, providing conditions conducive to cheatgrass invasion. Once established, cheatgrass alters fire regimes by creating continuous fine fuels that promote frequent, high severity fires (D'Antonio and Vitousek 1992). Sagebrush and other plants that are slow to recolonize following fire, like bitter brush and mountain mahogany, decline with increased fire frequencies. The invasion of cheatgrass has contributed to the conversion of thousands of hectares of sagebrush, bitter brush, and curl leaf mountain mahogany shrublands in the region to low diversity annual grasslands that do not provide habitat for native wildlife (Henstrom et al. 2002, Schaefer et al. 2003, Young 2000).

Western Juniper Woodlands

Western juniper woodlands occupy over 11% of the Modoc Plateau and Northwestern Basin and Range, representing the highest concentration of western juniper in California. It is found between 1,022 and 2,289 m in elevation, extending into sagebrush steppe vegetation on more mesic sites at the lower elevations, and into pine woodlands at higher elevations where it may co-occur with California black oak. This is the only location these two

on most soil types. Big sagebrush, ashy ryegrass, and bluebunch wheat grass are associated with juniper on deep alluvial soils. Wyoming big sagebrush, low sagebrush, and Thurber's needlegrass grow with juniper on shallow soils at lower elevations. Over the course of stand development, the interspaces between western juniper trees become completely occupied by spreading lateral roots, canopy cover increases, and shrub and herbaceous vegetation rapidly decline (Miller et al. 2000). Western juniper woodlands eventually become closed stands with virtually no understory species. These dense stands are also prone to high severity, stand-replacing fire.

Above: Cheatgrass invasion promotes frequent fires that eliminate woody shrub species. PETER WEISBERG

Below: Western juniper with low sagebrush on shallow soils. ALLISON SANGER

Below right: Western juniper and curl leaf mountain mahogany association on the first ridge of Eagle Lake near Fredonyer Peak. ROBERT WICK

species co-occur in the Great Basin. Other common associates include curl leaf mountain mahogany, rubber rabbitbrush (*Ericameria nauseosa*) and bitter brush.

Western juniper woodlands are often associated with volcanic soils, although they are not restricted to this substrate. In areas that are protected from fire, western juniper trees can live to be hundreds of years old. Western juniper woodlands in early succession are rich in broadleaf herbaceous species and perennial grasses. Squirreltail (*Elymus elymoides*) and curly blue grass (*Poa secunda*) typically co-occur with western juniper

Prior to the mid-1800s, juniper primarily occurred on dry, rocky substrates (West 1983). Since that time western juniper has undergone rapid expansion in the region, particularly in areas with deeper, more productive soils (Miller and Rose 1999). It is estimated that the extent of western juniper has increased ten-fold in the past 130 years and now covers more than one million hectares of northeastern California (CDFG 2006, Miller 2006). This expansion

likely results from a combination of grazing, fire exclusion, and climate change (Miller and Tausch 2001). Livestock grazing in the late 1800s and early 1900s consumed fine fuels, decreasing fire frequency and reducing competition from herbaceous species. The reduced fire frequency allowed western juniper to expand its range into sagebrush, bitter brush, curl leaf mountain mahogany, riparian, and aspen plant communities (Miller and Rose 1999). The consequences of western juniper invasion include the loss of native shrub and herbaceous plant cover and diversity, shifts to a high severity fire regime, reduced forage for wildlife and livestock, and decreased water availability (Miller et al. 2000, Miller et al. 2005).

Oak and Pine Woodlands

Mixed woodlands containing both oak and pine species are rare in the Modoc Plateau and Northwestern Basin and Range, as they are found at higher elevations where the climate is cooler and wetter than areas dominated by western juniper woodlands and sagebrush steppe. Oak and pine woodland alliances occur primarily in the western portion of the region in the Big Valley and Adin Mountains, but they are also scattered in the Devil's Garden area and the lower slopes of the Warner Mountains. Hardwood species are relatively rare here due to extreme cold temperatures and aridity. Existing stands may be remnants from a time when the regional climate was characterized by warmer minimum temperatures, higher precipitation, and lower evapotranspiration.

Alliances present here include California black oak forest, at its furthest northeast distribution, and Oregon white oak (*Quercus garryana*) woodland, at its most interior distribution in California. Oregon white oak is adapted to a relatively high frequency, low intensity fire regime. It is possible that Native American burning historically maintained many of these stands. Oregon white oak is found only in the Big Valley Mountain area, while California black oak occurs further east. Ponderosa pine (*Pinus ponderosa*) and Jeffrey pine (*P. jeffreyi*) co-occur with oaks on sites with deep, well-drained soils. Ponderosa pine is the most common pine in this type, but Jeffrey pine is

Above: Encroachment of western juniper trees on low sagebrush scrub with curly blue grass and Idaho fescue. The western juniper trees seen here in the Fitzhugh Creek Wildlife Area are all less than 15 years old.

Below: Large old western juniper trees can often be found on rocky outcrops where they have been protected from fire for hundreds of years.

Todd Keeler-Wolf photos

Top: Oregon white oak and ponderosa pine woodland.
Craig Odegard

Above: Oregon white oak woodland with ponderosa pine and western juniper.
Mike Dolan

also found in the southern part of the region. Jeffrey pine occurs on colder, higher, steeper and less productive sites than does ponderosa pine. Incense cedar (*Calocedrus decurrens*) is also present in this zone, particularly on wetter sites with deeper soils. Bitter brush is an understory dominant in these forests, as well as curl leaf mountain mahogany, Utah service-berry (*Amelanchier utahensis*), choke cherry (*Prunus virginiana*) and green leaf manzanita (*Arctostaphylos patula*) (Riegel et al. 2006). Grasses and graminoids are also common in these woodlands, including Idaho fescue, squirreltail, and Ross' sedge (*Carex rossii*). Mule ears (*Wyethia mollis*) is a wide-spread herb that may indicate historic heavy grazing.

Montane Forests

Higher in elevation than the oak and pine woodlands, montane forests in the Modoc Plateau and Northwestern Basin and Range differ from those in other mountain ranges in California. Typical Sierran conifers, including red fir (*Abies magnifica*), sugar pine (*Pinus lambertiana*), mountain hemlock (*Tsuga mertensiana*), and Douglas fir (*Pseudotsuga menziesii*) are largely absent (Griffin and Critchfield 1972). White fir (*Abies concolor*) is the indicator species for the mid and upper montane forests in the region (Riegel et al. 2006).

The mid-montane zone occurs in the Big Valley Mountains, the northern part of the Devil's Garden, and in the Warner Mountains. This zone is colder and wetter than the oak and pine woodlands, and supports white fir forests in pure stands or mixed with ponderosa and Jeffrey pines. Common shrub species in this zone include bitter brush, mountain big sagebrush, low sagebrush, bitter cherry (*Prunus emarginata*), choke cherry, Modoc plum (*Prunus subcordata*), Utah serviceberry, roundleaf snowberry (*Symphoricarpos rotundifolius*), green leaf manzanita, and tobacco brush (*Ceanothus velutinus*). Curl leaf mountain mahogany is also common on sites with shallow, rocky soils. Common grasses include squirreltail, western needlegrass (*Achnatherum occidentale*), Wheeler's bluegrass (*Poa wheeleri*), and Orcutt brome (*Bromus orcuttianus*).

At higher elevations, upper montane forests often occur in more open, discontinuous stands intermixed with herb and shrub dominated communities,

Top: A mosaic of upper montane communities including low sagebrush in the Warner Mountains.
Mike Dolan

Below: Mid-elevations of the Warner Mountains are blanketed in white fir.
Michael E. Kauffmann

Top: Looking south at the Summit Ridge of Warner Mountains with the steep glaciated side to the east and the more gradual gentle slope to the west.
Rachel Simons

Above: Washoe pine in the upper montane forest.
Forest Gauna

meadows, montane riparian scrub, and aspen. White fir occurs in pure stands and mixed stands with western white pine (*Pinus monticola*), lodgepole pine (*Pinus contorta* ssp. *murrayana*), aspen, ponderosa pine, Jeffrey pine, Washoe pine (*Pinus ponderosa* var. *washoensis*), and whitebark pine (*Pinus albicaulis*) at the upper elevations of this zone (Riegel et al. 1990, Smith 1994). Western juniper woodland can also be found in the upper montane on exposed sites with shallow soils. Common shrubs include mountain big sagebrush, low sagebrush, creeping snowberry (*Symphoricarpos mollis*), sticky currant (*Ribes viscosissimum*), bitter cherry, tobacco brush, curl leaf mountain mahogany, Scouler's willow (*Salix scouleriana*), and pinemat manzanita (*Arctostaphylos nevadensis*) (Riegel et al. 1990, Smith and Davidson 2003). Washoe pines, a high elevation variety of ponderosa pine, can tolerate harsher environmental conditions than ponderosa pine, and often occurs in open woodlands with grassy or shrubby understories. Washoe pine woodlands are largely restricted to this region and are an important member of upper montane communities, comprising up to 70% of stands in some areas.

Vegetation structure and composition in montane forests have been altered by a combination of fire suppression and historic grazing. There has been a shift in the last century to less pine and greater white fir dominance, as well as higher stand densities (Smith 1994). Like most forests throughout California, these areas are now prone to larger, higher severity fires and increased susceptibility to drought, insects, and disease.

Subalpine

Subalpine vegetation is relatively uncommon in the Modoc Plateau and Northwestern Basin and Range; it is confined to the Warner Mountains. The range is very steep and glaciated on the east side, but the west-facing slopes are gentle. Whitebark pine, an important

species in this zone, was recently declared a candidate for federal listing under the Endangered Species Act because of threats from climate change, fire suppression, mountain pine beetle (*Dendroctonus ponderosae*), and white pine blister rust (*Cronartium ribicola*) throughout much of its range.Whitebark pine forests are typically found in cold, windy, high-elevation sites that are inhospitable to many other conifer species. Whitebark pine is considered a keystone species because it regulates runoff by slowing the progress of snowmelt, reduces soil erosion by initiating early succession after fires and other disturbances, and provides seeds that are a high-energy food source for mammals and birds, such as Clark's nutcrackers (*Nucifraga columbiana*). Curl leaf mountain mahogany scrub occasionally occurs in isolated stands on exposed upper slopes. Common herbaceous species found in whitebark pine forests include western needlegrass, pussypaws (*Cistanthe umbellata*), Wheeler's bluegrass, slender penstemon (*Penstemon gracilentus*), prickly sandwort (*Arenaria aculeata*), Davidson's penstemon (*Penstemon davidsonii*), and spreading phlox (*Phlox diffusa*) (Riegel et al. 2006). Patches of montane riparian scrub, black cottonwood (*Populus trichocarpa*) forest, and subalpine sagebrush are also scattered throughout the subalpine (Riegel et

Above: Curl leaf mountain mahogany scrub. MIKE DOLAN

Below: Whitebark pine forest showing mortality caused by mountain pine beetles. RACHEL SIMONS

Top: The Timbered Crater Wilderness Study Area supports the largest known population of Baker cypress.

Above: Baker cypress seedlings after a fire at Timbered Crater Wilderness Study Area.
Kyle Merriam photos

al. 1990). Alpine plants such as alpine spring beauty (*Claytonia megarhiza*), alpine ivesia (*Ivesia gordonii*), mountain sorrel (*Oxyria digyna*), and creeping sibbaldia (*Sibbaldia procumbens*) occur at the highest elevations.

Baker Cypress

The Modoc Plateau supports the largest population and type locality of Baker cypress, a rare species known from only 11 widely scattered locations across the northern Sierra Nevada, Cascade, and Siskiyou Mountains. Located within the 7,284-hectare Timbered Crater Wilderness Study Area, this Baker cypress population is scattered across a large volcanic crater and surrounding lava flows on the bioregion's western edge at elevations from 1,067 to 1,219 m. This population occupies approximately 3,000 ha, primarily growing in monospecific stands. Other species occurring in Baker cypress stands include knobcone pine, ponderosa pine, and Oregon white oak. The understory is dominated by green leaf manzanita. Additional shrubs present include bitter brush, wedge leaf ceanothus or buck brush (*Ceanothus cuneatus*), deer brush (*C. integerrimus*), curl leaf mountain mahogany, oceanspray (*Holodiscus discolor*), and western redbud (*Cercis occidentalis*).

This mix of Californian and Great Basin species illustrates the transitional nature of the flora in this part of the region. Understory herbs are diverse, and although they represent a tiny amount of herbaceous cover, several annual species are abundant and widespread. These include opposite-leaved tarweed (*Hemizonella minima*), threadstem fireweed (*Epilobium minutum*), Pacific fescue (*Vulpia microstachys*), slender popcorn flower (*Plagiobothrys tenellus*), Douglas' stitchwort (*Minuartia douglasii*), and maiden blue eyed Mary (*Collinsia parviflora*). Perennial species such as Ross' sedge, scabland Penstemon (*Penstemon deustus* var. *pedicellatus*), threetooth horkelia (*Horkelia tridentata*), and buckwheats are also widespread but scattered.

Fire is the only effective process that opens the serotinous cones of Baker cypress and creates conditions for successful regeneration. Due to its remote location and inaccessibility, fires have not been suppressed within the Timbered Crater area and fires have been recorded here in 1910, 1977, 1992, and 2000. Unlike many other Baker cypress stands, which are largely decadent and lack any signs of reproduction, the Baker cypress population here contains

a range of age classes, with cored trees ranging from 25–125 years in age and numerous seedlings evident in recently burned areas. The Timbered Crater Baker cypress population is the healthiest and most robust population of Baker cypress known. The unproductive soils and remote location of this population have allowed it to flourish in the absence of competition and fire suppression or other detrimental management activities. Continued protection of this area and a policy of not suppressing fires here will likely maintain this impressive Baker cypress stand without any other management actions.

Vernal Pools

Vernal pools are seasonal wetlands that support diverse annual floras. Species adapted to different lengths of inundation, flowering sequentially as the pool dries in the spring, creating distinctive rings of flowers surrounding the pools. Northern basalt flow vernal pools are found in the region, classified as the Modoc Plateau vernal pool bioregion by Keeler-Wolf et al. (1998). Holland mapped 13,000 ha of vernal pools on the Modoc National Forest alone (Holland 2006), primarily in the Devil's Garden area. These vernal pools are floristically somewhat different from those in other

Top: Vernal pool filled by winter snowfall. Meredith Gosejohan

Above: Vernal pool in spring on the Devil's Garden. Forest Gauna

vernal pool regions in California, because they experience a much colder climate, and occupy up to a thousand hectares in some areas. The surrounding vegetation types are unique for California vernal pools, including silver sagebrush (*Artemisia cana*) scrub, low sagebrush scrub, big sagebrush shrubland, as well as Jeffrey pine forest and western juniper woodland. Several federally listed vernal pool plant species occur here, including slender orcutt grass (*Orcuttia tenuis*) and Greene's tuctoria (*Tuctoria greenei*). These species have disjunct distributions, with most occurrences found in the Great Valley of California, and the remaining known populations found on the Modoc Plateau. Other common vernal pool species in the bioregion include beethistle (*Eryngium articulatum*), needle leaf navarretia (*Navarretia intertexta*), milkwort knotweed (*Polygonum polygaloides*), Bacigalupi's downingia (*Downingia bacigalupii*) and dwarf wooly-marbles (*Psilocarphus brevissimus*).

Unlike vernal pool habitats in the rest of California, where it is estimated that 80-90% of the vernal pools historically present have been lost (Holland 1998), most vernal pools in this region are on federally managed land and are protected from agricultural conversion or commercial development. The vernal pools have still been inadvertently degraded, particularly through alteration of vernal pool hydrology (USFS and BLM 2012). Holland (2006) found that over one third of the 660 vernal pools mapped on the Modoc

*Top: Tricolored monkeyflower (*Mimulus tricolor*) with Bacigalupi's downigia at Green Place vernal pool.*
Mike Dolan

Middle: Bacigalupi's downingia.

Bottom: Slender orcutt grass at large Coyote vernal pool after a dry winter with some late rains.
Don Lepley photos

National Forest had some hydrologic degradation from ditching and damming.

Meadows and Riparian Areas

Riparian areas are defined as bordering streams, while meadows can occur with or without streams. Meadows and riparian areas support diverse and rapidly growing vegetation in summer. Meadows are found scattered throughout the region at elevations ranging from 1,010 to 2,660 m, on fine-textured, moist, nutrient-rich soils. Montane meadows are found in the Warner and Big Valley Mountains, including vegetation alliances such as Nebraska sedge (*Carex nebrascensis*) meadows, small-fruited sedge (*Carex integra*) meadows, short-beaked sedge (*Carex simulata*) swards, tufted hair grass (*Deschampsia cespitosa*) meadows, and white corn lily (*Veratrum californicum*) patches. These montane meadows are often heavily utilized by cattle. Lower elevation meadows are found scattered in basins, plains, and plateaus throughout

Above: Dry meadow including flowering yampah (Perideridia *sp.*).
Mike Dolan

Bottom: White corn lily in a montane meadow.
Dave Weixelman.

the region, such as at Steele Swamp, Jess Valley, Ash Meadows, and along the Pit River. These lower elevation wet meadows include Baltic rush (*Juncus arcticus* ssp. *littoralis*) marshes and Nebraska sedge meadows, and dry meadows supporting upland species such as Idaho fescue. Meadow complexes have been protected on federal lands at higher elevations, and at lower elevations on state refuges and wildlife areas such as the Willow Creek Wildlife Area. Historic grazing practices, up until approximately the 1980s, resulted in significant degradation of many meadows. Since that time period,

Top: Red osier (Cornus sericea) *thickets and sedge meadows in the riparian area along Fitzhugh Creek.*

Left: Riparian and fen vegetation at Dismal Swamp in the Warner Mountains, including resin birch and diverse sedge communities.
TODD KEELER-WOLF PHOTOS

USFS monitoring has shown that meadow health has an upward trend on USFS lands. Many lower elevation meadows have been hydrologically altered by irrigation ditches and are harvested as hay or irrigated for pasture. A number of organizations, such as the Pit River Watershed Alliance, have emphasized riparian and meadow restoration on private lands.

Fens—groundwater-fed organic rich wetland ecosystems that support high levels of plant and animal biodiversity—are also found in meadows in the region, primarily in the Devil's Garden area and the Warner Mountains. Fens can sequester carbon and store peat for thousands of years, making them invaluable for paleoclimatic reconstructions and important indicators for climate change. Important fen indicator species include mosses such as three-ranked humpmoss (*Meesia triquetra*), broad-nerved humpmoss (*Meesia uliginosa*), and sphagnum moss (*Sphagnum* spp.). Fens also support highly diverse sedge (*Carex* spp.) communities.

Riparian vegetation occurs along streams and can be found scattered throughout the region, including alpine and subalpine habitats in the Warner and Big Valley Mountains and along stream terraces and rivers in lower elevation valleys and plains, including sections of the Pit and Susanville Rivers and their tributaries, such as Ash Creek and Willow Creek. Primary riparian vegetation types identified in the region by Smith (1998) include: aspen mixed with conifers; deciduous riparian forests dominated by Oregon ash (*Fraxinus latifolia*) or black cottonwood; willow (*Salix* spp.) dominated types; types lacking willows and characterized by species such as Pacific dogwood (*Cornus nuttallii*), silver sagebrush and Utah

Above: Aspen groves in the Warner Mountains.
FOREST GAUNA

Top: Isolated aspen groves can occur in basalt outcrops.
Forest Gauna

Above: Close up of aspen grove in the Warner Mountains.
Todd Keeler-Wolf

serviceberry; and grass and forb dominated types with species composition varying by elevation. Higher elevation riparian alliances include resin birch (*Betula glandulosa*) thickets and Geyer willow (*Salix geyeriana*) thickets. Lower elevation riparian alliances include water birch (*Betula occidentalis*) thickets, sandbar willow (*Salix exigua*) thickets, and arroyo willow (*Salix lasiolepis*) thickets. Although some of these riparian corridors are protected in state wildlife areas, much of the historic lower elevation riparian habitat has been lost to agricultural conversion.

Aspen

Aspen groves occupy just over 1,000 hectares in Northeastern California. They can be found at a range of elevations, from 1,264 to 2,639 m, primarily concentrated along stream banks, stream terraces, and toe slopes in the Warner Mountains. Aspen is otherwise uncommon north of the Sierra Nevada range in California. In the Warner Mountains, aspen groves are typically mixed stands bordering Jeffrey pine, white fir, and Washoe pine forests, but can also form pure stands. Outside of the Warner Mountains, small aspen groves can occur on basalt outcrops and in isolated pockets in sagebrush steppe and grassland areas such as the Clear Lake Hills. Aspen groves are an important component of the region because they support high levels of understory biodiversity, provide habitat for wildlife, and are an indicator of ecological integrity. Aspen groves are declining as a result of fire suppression and overgrazing across much of the west. A comparison of aerial photographs from 1946 and 1994 in the Warner Mountains found that the total area occupied by aspen declined

by 24% during this 48-year period, and that modern aspen groves are smaller and more fragmented (Di Orio et al. 2005).

Marshes and Emergent Wetlands

Marshes and emergent wetlands are scattered throughout the region. These areas represent remnants of formerly vast wetlands. Most of the remaining wetlands and marshes are managed and maintained as wildlife areas or refuges by CDFW, BLM, and USFWS. Marsh and emergent wetlands include the cattail (*Typha* spp.) marsh and bulrush (*Schoenoplectus* spp.)

Above: Cattail marsh at Tule Lake.
John O. Sawyer

Right: Wetlands on the Lower Klamath National Wildlife Refuge provide habitat for many wildlife species.
U.S. Fish and Wildlife Service

Top: Greater sage-grouse.
Robert Wick

Above: Restoration effort of sagebrush steppe in the Modoc National Forest where western junipers were cut with chainsaws and left on-site.
SageSTEP

marsh alliances. These species can tolerate both brackish and freshwater, and are found along streams, pond, lakes, sloughs, and swamps. Bulrush marsh alliances are found at Clear Lake, Lower Klamath Lake, Tule Lake, Goose Lake, and in the Willow Creek Wildlife Area near Honey Lake. Cattail marshes are found throughout the region except at higher elevation sites. Extensive emergent wetlands and marshes are protected on the Klamath Basin National Wildlife Refuge Complex, the Modoc National Wildlife Refuge, and the Ash Creek Wildlife Area.

Marshes and emergent wetlands provide critical habitat for migratory and breeding habitat for hundreds of species of waterfowl, white pelicans, double-crested cormorants (*Phalacrocorax auritus*), gulls (*Laridae* spp.) and numerous marsh and upland birds such as herons (*Ardeidae* spp.), and sandhill cranes. These extensive wetlands attract peak fall concentrations of over six million waterfowl. Unfortunately, like many wetlands in the region, the Klamath Basin has lost 80% of its original wetlands. Threats to remaining wetlands include degraded water quality and alterations to water quantity.

Wildlife: Greater Sage-Grouse

The greater sage-grouse (*Centrocercus urophasianus*) is completely dependent on sagebrush steppe vegetation for both foraging and nesting habitat. Greater sage-grouse populations in the Modoc Plateau represent the western extent of the species' range. Seasonally, greater sage-grouse gather in specific breeding display sites called leks that are used year after year, unless the habitat is altered or destroyed. Each lek may host 10–100 or more male birds that exhibit brilliant strutting displays to attract females. As a result of habitat degradation by livestock grazing, the invasion of exotic grasses, increased fire frequency, the expansion of western juniper, and the construction of roads, fences, and utility poles, greater sage-grouse populations have declined sharply in the past several decades with many populations decreasing by 45–80% since 1950. In the Modoc Plateau and Northwestern Basin and Range, greater sage-grouse populations have declined from 45 active leks in the 1960s to only one lek with fewer than 100 individuals remaining in 2005 (Hall 1995).

The greater sage-grouse is the focus of numerous efforts to restore sagebrush habitats across the West, including the Greater Sage-grouse Conservation Plan for Nevada and Eastern California developed by Nevada Department of

Wildlife, Greater Sage-grouse and Sage-Steppe Ecosystems Management Guidelines developed by BLM, the Sage Grouse Initiative developed by Natural Resource Conservation Service, and the Sage-Steppe Ecosystem Restoration Strategy developed by USFS and BLM. Restoration of the sagebrush steppe will benefit numerous other sagebrush dependent species such as sagebrush sparrow (*Artemisiospiza nevadensis*), sage thrasher (*Oreoscoptes montanus*), Brewer's sparrow (*Spizella breweri*), pygmy rabbit (*Brachylagus idahoensis*), and pronghorn antelope (*Antilocapra americana*).

Conservation Efforts

In addition to the greater sage-grouse conservation efforts described above, private land owners, state and federal land management agencies, resource conservation districts, watershed groups, and fishing and hunting organizations are involved in a number of stream, riparian, wetland, and upland restoration and conservation projects. The Cooperative Sagebrush Steppe Restoration Initiative, a coordinated effort of BLM and USFS, developed a suite of landscape-level treatments to reestablish sagebrush steppe plant communities across much of the region. Central Modoc Resource Conservation District and Pit River Watershed Alliance are working with private landowners to develop and implement stream restoration projects. USFS has fenced some streams to protect the endangered Modoc sucker (*Catostomus microps*) and other species, and have implemented rotational grazing systems to restore riparian habitats. USFS and BLM have conducted wild horse round-ups to reduce damage from feral horse overgrazing. The Goose Lake Fishes Working Group has been working to restore native fishes of the Goose Lake Basin. These are just a few of the conservation efforts underway to restore the Modoc Plateau and Northwestern Basin and Range.

Kyle Merriam

Low sagebrush scrub on Shaffer Mountain, northeast of Susanville.
Robert Wick

CHAPTER FIVE

Mono and Southeastern Great Basin

The eastern fringe of California slices off a thin strand at the edge of a vast interior biome, the Great Basin. Often characterized as an immense and homogeneous sagebrush (*Artemisia* sp.) sea, this region in fact encompasses great topographic, geologic, climatic, and vegetative diversity, haunting in expansiveness of landscape, surprising in richness of hidden canyons and wetlands. Long lines of basin and range draw the eye outward to where land meets sky; wave after wave of mountain ranges pounding the sage surf. If only a slice, California is fortunate to have what it does of this Great Basin.

The crest of the Sierra Nevada defines the western hydrologic edge of the Great Basin, within which waters drain into interior basins. The entirety of the Sierra Nevada was built by similar geologic forces that created the Great Basin. The narrative for vegetation in this chapter starts with the lower montane (~2,500 m at the latitude of Mono Lake) and basin bottom environments of the eastern Sierra Nevada, and goes on to embrace the entire elevation gradients from basin to summits of the mountain ranges eastward to the California-Nevada state line. Their names sing of romance

Above: The crest of the Sierra Nevada defines the western edge of the Great Basin in central eastern California. The Mono Basin, with its crown jewel Mono Lake, exemplifies the character of the California part of this province, with expansive cold-adapted sagebrush steppe intermixed with hardy forbs and grasses.
ROBERT WICK

Opposite: A winter evening along Hot Creek, with rubber rabbitbrush and sagebrush spread over rolling hills of the Long Valley Caldera.
JEFF BISBEE

The great height of the Sierra Nevada (background) creates a rainshadow effect on lands to the east, leaving the Great Basin much drier than the California floristic province. Both the basin floors, such as the Owens Valley south of Bishop, and the higher slopes, such as the White and Inyo Mountains, maintain an arid character, and support dry-adapted plants such as cottontop cactus (Echinocactus polycephalus).
Robert Wick

and adventure: Sweetwater, Bodie, Glass, White, Inyo, Grapevine, Coso, Panamint, Amargosa, Argus, Last Chance, Funeral. Even the most recently named range of the Great Basin, the Zunamed Mountains (east of Ebbetts Pass), evokes mystery. Summits extend from 1,800 m in the smaller ranges to 4,342 m atop White Mountain Peak in the White Mountains, California's third highest mountain. Basin bottom elevations generally run higher in the north (above 1,800 m), with southern basins averaging about 1000 m. California's oldest and most important land-survey reference, the Mt. Diablo Baseline (~38° N), runs through the Mono Basin, dividing the region into north and south parts. Taken together, the landscape of this chapter, hereafter referred to as the California Great Basin, includes nearly 1.9 million ha (4.7 million ac) or 4.6% of California's land area.

The dominant climate of the California Great Basin is cool and dry (annual precipitation ranges from 52 cm in northern uplands to 11 cm in southern basins), although low basins bordering the Mojave province in the south are warmer and drier. Farther interior in the Great Basin, continental climates drive very cold winters and hot summers; the California borderlands are mostly buffered from these extremes. Despite semi-aridity, the region's climate is dominantly influenced by winter storm tracks and high pressure originating from the Pacific Ocean and then moving eastward over the Sierra Nevada. Orographic effects catalyze waves of precipitation on the west slopes, contrasting with progressively stronger rain-shadows in the lee of mountain crests. The impressive height and location of the Sierra Nevada—as the first major range downwind from Pacific storms—capture and hold precipitation in deep snowpacks. Precipitation declines down the mountains and farther east (e.g., 70 cm annually at Tioga Pass, 35 cm in the western Mono Basin, and 13 cm in the eastern Mono Basin). Snow falls on the high peaks of the adjacent California Great Basin ranges, but in far lesser amounts than on the Sierra Nevada, leaving the lee mountains to be among the driest in the Great Basin for their latitudes.

Secondary climatic influences vary with geography and season. A common winter pattern is the Great Basin High that develops over cold central Nevada. This blocks and diverts Pacific storm tracks northward, leaving the California Great Basin dry, clear, and cold for long periods. In summer, the southern California Great Basin often receives monsoon influence from the Gulf, and convectional forces deliver an artillery of thunderstorms to the uplands throughout the region. Both result in minor amounts of summer precipitation in the mountains. Occasionally, summer storms stall over the high slopes, driving impressive flash floods down mountain canyons.

These environmental forces also shape the region's vegetation. As elsewhere in California, gradients of climate create repeating changes in vegetation across the California Great Basin. Elevation is commonly associated with changes in climate and vegetation—temperatures being on average colder with increasing altitude. This is not always the case, however, as inversions and other types of cold-air drainage—that is, where temperatures decrease as elevations decline—are common. In depressions, valley bottoms, and basins, plant communities must be able to tolerate super-cold temperatures that occasionally develop. Latitude and longitude also correspond to gradients in climate and vegetation response across the region, especially in regard to rain shadow and eastward effects. The degrees of latitude (> 4) spanned by the California Great Basin weave a tapestry of vegetation diversity from north to south. Northern communities align more with cold shrub- and forest types, while southern communities border warm deserts and support dry scrublands. Finally, changes in substrate create their own repeating patterns of vegetation. In many basins, for instance, gradual changes in sediment size and salinity (often the legacy of ancient

Top: Summer convectional storms blow up frequently in the California Great Basin province, delivering welcome rainbursts during the otherwise long, dry summers, yet also igniting fires in the woodlands and sagebrush steppe.
Julie M. Evens

Above: Big sagebrush in the snow-covered Mono Basin.
Connie Millar

Above: The Racetrack in Death Valley National Park is one of the old lake beds, or playas, occupying isolated basins of the Great Basin. Todd Keeler-Wolf

*Top right: Royal penstemon (*Penstemon speciosus*) here on Freel Peak in the Carson Range, joins other showy plants that provide a riot of color to the alpine zone of the California Great Basin.*

Opposite, top: The botanically diverse Sweetwater Mountains, north of Bridgeport, contain a suite of endemic plants, as well as species hailing from the Sierra Nevada to the west and Great Basin provinces to the east. The former reflects upland connections among the ranges, which function as biogeographic corridors.

Opposite, center: Pacific alpine gold dances and shimmers on the high slopes of Ferris Canyon in the Sweetwater Mountains. Connie Millar photos

terminal lakes) can create concentric rings of vegetation. Elsewhere, exposures of substrates that impose nutrient deficiencies and toxicities to plants, such as dolomites and hydrothermally altered soils, define vegetation mosaics that can have extremely abrupt boundaries.

The California Great Basin abuts the Sierra Nevada flank at mid-low montane elevations along a broad north-south zone. As described in the Sierra Nevada chapter, bands of elevationally-adapted vegetation rise in altitude as latitude decreases, making the boundary of the Sierra Nevada and California Great Basin regions higher in the south than north. In the north, the smaller ranges east of the Sierra, such as the Sweetwater, Zunamed, Bodie Hills, and Glass Mountains, do not show strong latitudinal effect, but reflect east-west rain shadows compounded by biogeographic connections that their west slopes have with the Sierra Nevada. The latter provide access of Sierra Nevada species into these adjacent ranges, and also tend to draw storm tracks from the Sierra Nevada eastward into adjacent uplands. By contrast, the high and expansive White-Inyo Mountain chain is isolated from the Sierra Nevada by the broad, low, and deep Owens Valley, so it reflects a more interior character. Smaller ranges in the south, such as Coso, Last Chance, Funeral, and Grapevine, have vegetation affinities with southern Nevada and Mojave communities.

Alpine Herbaceous Communities

Alpine communities—those that occur above climatic treeline—clothe those mountains of the California Great Basin that rise above 3,300 m in the north and above 3,475 m in the south. These cold, arid environments are dominated primarily by herbaceous species, including many bunch grasses and sedges, although some hardy shrubs dot the landscape, especially in sheltered locations. Given the harshness of the environment, alpine species are specifically adapted to these conditions and many occur only in these elevations. Species that are widespread across elevational gradients and creep into the alpine zone are discussed later. Topography, snow accumulation pattern, and soil depth, impose a patchwork pattern of deep brilliant greens, silvery grays, and cheery yellows and pinks to the patterns of vegetation. Soggy tussock meadows alternate with exposed gravelly alpine hills, steep talus slopes, and outcrops dotted with occasional brilliant blues of sky pilot (*Polemonium eximium*), cushion buckwheat (*Eriogonum ovalifolium*), or cut-leaved daisy (*Erigeron compositus*).

In rocky fell fields of the highest ranges, such as the White and Inyo Mountains and the Sweetwater Mountains, the charming fell-fields with California heath-goldenrod and Pacific alpine gold (*Ericameria discoidea–Hulsea algida*) sparsely vegetated alliance greet the high mountain traveler with flowers beaming vibrant hues of yellow and gold. Both species scent the thin air with distinctive and pungent aromas. California heath-goldenrod is a multi-stemmed subshrub with dense and delicate inflorescences, while Pacific alpine gold stands tall on single, erect stems. The sturdy composite floral head of the latter is showy and glows a rich, incandescent orange. While California heath-goldenrod grows commonly across rocky slopes, Pacific alpine gold seems to prefer the windiest ridge crests and most exposed alpine summits, often found grouped in cheery clusters seeming impossibly out of place amidst the harsh environment. Common co-occurring species include several buckwheats (*Eriogonum* spp.), in particular the widespread cushion buckwheat, graceful blue-flowered silverleaf phacelia (*Phacelia hastata*), delicate King's sandwort (*Arenaria kingii* ssp. *kingii*), the common sub-shrub granite prickly phlox (*Linanthus pungens*), and perennial bunch grasses such as pine needlegrass (*Achnatherum pinetorum*), squirreltail (*Elymus elymoides*), prairie junegrass (*Koeleria macrantha*), and Wheeler's bluegrass (*Poa wheeleri*).

The expansive lunaresque alpine slopes and fell fields of the Sweetwater Mountains host diverse alpine species, including many rare species endemic to California. A characteristic alliance is the cushion phlox (*Phlox pulvinata*) fell-fields, with spiny milkvetch (*Astragalus kentrophyta*), Gordon's ivesia (*Ivesia gordonii*), and white ballhead gilia (*Ipomopsis congesta*). Cushion phlox has

*Left: Hardy plants adapted to cold, windy, and arid conditions of Great Basin alpine ecosystems grow in the high elevations. Here rayless shaggy fleabane (*Erigeron aphanactis*) and cushion buckwheat provide bursts of color against the light soils of the Sweetwater Mountains.*
Connie Millar

Above: Swathes of cushion phlox alliance along Wheeler Peak in the Sweetwater Mountains.

Opposite, top: Tufted hair grass alpine communities on Wheeler Peak of the Sweetwater Mountains often contain scattered krummholz conifers, such as whitebark pine, near upper treeline.

Opposite, bottom: Low sage-steppe communities of little sagebrush and Rothrock's sagebrush dominate alpine slopes of Great Basin mountains, here in the Cottonwood Basin of the White Mountains.
Connie Millar photos

an indescribably sweet aroma when it flowers in early summer, blanketing dry, upper slopes of the Sweetwater Mountains visually and aromatically with intoxicating beauty. Squirreltail is common in this alliance, and throughout California, where it is widespread in diverse habitats.

Tufted hair grass (*Deschampsia cespitosa*) meadows are common from sea level to high mountains, but in the California Great Basin they are restricted to the alpine zone. In the White Mountains and meadows of the Sweetwater Mountains this bunchgrass grows with forbs such as fireweeds (*Epilobium* spp.), yampah (*Perideridia parishii*), little elephant heads (*Pedicularis attollens*), and others varying by locale. Several sedges often co-occur in these meadows, including Nebraska sedge (*Carex nebrascensis*), dark alpine sedge (*C. subnigricans*), and beaked sedge (*C. utriculata*).

Alpine Shrublands

Low sage-steppe communities are widespread in the alpine zones of the California Great Basin and extend across high-elevation ridges in the White Mountains and upper slopes of the Inyo, Cottonwood, and Panamint Mountains. Little sagebrush (*Artemisia arbuscula* ssp. *arbuscula*) and Rothrock's sagebrush (*A. rothrockii*) dominate alliances of their names. They occur in areas of shallow soils, often at margins of high-elevation meadows; Rothrock's sagebrush scrub

alliance grows in soils derived from a variety of bedrock types, from granitic to carbonate-rich dolomites and limestone, as in the White and Inyo Mountains. While technically a tree species, extensive thickets of stunted (krummholz) conifers grow in the alpine zone. These include whitebark pine (*Pinus albicaulis*), the queen of krummholzing conifers, which extends above treeline in the Sweetwater and Glass Mountains, as well as clusters of wind-ravaged limber pine (*Pinus flexilis*) in the Bodie Hills and the Sweetwater, Glass, White-Inyo, and Last Chance Mountains. Bristlecone pine (*Pinus longaeva*), the other high-elevation conifer, rarely takes on a stunted form above treeline. Low thickets of shrubby aspens (*Populus tremuloides*), glowing

Top: Whitebark pine krummholz matts can extend for more than 300 m in a band above upright trees at the thermal treeline, as in Sweetwater Canyon in the Sweetwater Mountains.

Above: Limber pines growing near treeline, here in Sweetwater Canyon on the east slopes of the Sweetwater Mountains, respond to harsh conditions less by adopting krummholz form but more through stripbark growth and spiral grain, which gives trees a corkscrew appearance.
Connie Millar photos

golden in the fall, spread clonally up from subalpine meadows to surround talus slopes at even higher elevations than the bristlecone pines in McAfee Creek, Crooked Creek, and other canyons on the eastern slopes of the White Mountains.

Montane Forests

Below the alpine zone a kaleidoscope of forest, shrub, and herb communities paints the California Great Basin landscape, wherein plant communities follow preferences of topography, aspect, micro-climate, soils, competition, and gradients of moisture. Forests are the major architect of vegetation in the montane zones of the eastern Sierra Nevada and northern ranges, although they are often sparse, especially near upper and lower treelines. Comprising the greatest area of forest are conifer types, and among these, pines dominate. The most common subalpine conifer vegetation in mountain ranges of the California Great Basin is limber pine woodland, which occurs in the Bodie Hills, the Sweetwater, Glass, White-Inyo, Panamint, and Last Chance Mountains—and possibly high points elsewhere, including the Grapevine Mountains. In the eastern Sierra Nevada, limber pine occurs mostly above the elevation of the California Great Basin region, and is not documented north of Sonora Pass. Limber pine woodlands range from closed forests (less common) mixed with other species such as lodgepole pine (*Pinus contorta* ssp. *murrayana*), as in the Glass Mountains, to open sparse woodlands of scattered old-growth trees, as on the high slopes of the Sweetwater and White-Inyo Mountains. Long-lived (to 1,700 years), the limber pines' trunks often appear wizened as strips of bark commonly die back over time, and the deeply furrowed stems glow golden-red with age.

Limber pine grows with the iconic and photogenic bristlecone pine in the highest southern ranges of the region, the White-Inyo Mountains and the Panamint Range. Bristlecone pines are widely appreciated for their longevity, and individuals have been measured living more than 4,800 years. Tree-ring scientists have deciphered eleven millennia of climate records from annual growth rings by combining live and dead tree records. Like limber pine woodlands, bristlecone pine woodlands reach into the California Great Basin from their broader distribution outside California, which includes high peaks of southern-central Nevada, the Colorado plateau, and western Rocky Mountains (including related *Pinus aristata*). In the California Great Basin these forests are most often very open woodlands, with old-growth trees separated many meters apart. In middle elevations, younger stands approach closed-canopy conditions, but never so dense as in ranges east of the state line. In the White and Inyo Mountains, bristlecone pines cleanly define substrate transitions with their capacity to grow on dolomite and limestone, which exclude many other species. In areas where carbonate soils meet sandstones and shales, for instance, the contact is revealed as an abrupt boundary dividing open bristlecone pine woodland with little to no understory on one side and dense sagebrush steppe on the other.

Whitebark pine forests are the most common subalpine forest type in high elevations of the eastern Sierra Nevada,

Above: Aspen groves follow rocky stream courses and fringe wet meadows in Great Basin mountains, as in McGee Creek.
Jeff Bisbee

Below: Bristlecone pines in the White and Inyo Mountains rarely take shrubby stature even near treeline. Their capacity to endure for more than five millennia testifies to their endurance in the cold, arid environments, as here in the Inyo Mountains.
Robert Wick

Top: Bristlecone pines growing on dolomite soils near Patriarch Grove in the White Mountains.
Jeff Bisbee

Above: Whitebark pine in the California Great Basin grows scattered on granitic knobs along the lower slopes of the Sierra Nevada, as here below Mt Dunderberg.

Right: Whitebark pine depends on Clark's nutcracker to open cones and plant the seeds.
Connie Millar photos

but the occurrence of the species in California Great Basin elevations of the Sierra Nevada is limited to admixture in forests dominated by other conifers. In several of the northern ranges, whitebark pine forests are extensive—for instance in the Zunamed, Sweetwater, and Glass Mountains, they dominate the subalpine zones and form upper treeline. In those areas, whitebark pine forests comprise clusters of trees established from seed caches of Clark's Nutcracker (*Nucifraga columbiana*), a handsome relative of the crow that plays an essential, co-adapted role in dispersal of the pine seeds.

Mixed conifer forests typical of

mid-elevations on the western slope of the Sierra Nevada are less widespread along the eastern escarpment and their compositions there are simpler. In the zone of contact between the Sierra Nevada and California Great Basin regions, a patchwork of conifer forests includes alliances with varying dominances of lodgepole pine, white fir (*Abies concolor*), and Jeffrey pine (*Pinus jeffreyi*). Western white pine (*Pinus monticola*) and western juniper (*Juniperus occidentalis*) are occasional partners. These forests also grow at mid-to-upper elevations in the northern ranges of the California Great Basin, the Zunamed, Sweetwater, Bodie Hills, and Glass Mountains, where the intrepid botanist can find mountain hemlock (*Tsuga mertensiana*) as a rare component.

In these ranges, open Jeffrey pine forests are common, often monotypic, but in some cases intermixed with lodgepole pine, especially on dry, sterile, or fire-prone sites, and with white fir in more mesic locales. Jeffrey pine was prized for lumber during the early mining and settlement era of the region, so historic harvest levels were high and the largest remaining old-growth forests are in the Indiana Summit Research Natural Area, south of Mono Lake in the Glass Mountains. An interesting aspect of the Jeffrey pine forest in the southern Mono Basin are remnants of piagi rings, or trenches, surrounding the base of old-growth trees. These were dug by native Paiutes to harvest the large and nutritious larvae of the Pandora moth (*Coloradia pandora*), which undergoes irregular and episodic outbreaks in the Jeffrey pine forests. Sentinel Meadow Research Natural Area to the east along the crest of the Glass Mountains also contains fine stands of Jeffrey and lodgepole pines, mixed with ancient limber pines. The Volcanic Tablelands of Bishop tuff south of the

Top: Jeffrey pine occurs in riparian corridors of the lower eastern Sierra Nevada, here in Convict Canyon.
JEFF BISBEE

Above: Jeffrey pine in the Indiana Summit Research Natural Area, Mono Basin. This "Piagi tree" has trenches around its stem dug by early Paiute people to collect larvae from the Pandora moth.
CONNIE MILLAR

Above: Jeffrey pine woodlands near Swall Meadows above the Owens Valley often mix with big sagebrush steppe, where diverse herbs and small shrubs bloom, including lupine (Lupinus spp.)*, paintbrush* (Castilleja spp.)*, and sulphur buckwheat* (Eriogonum umbellatum)*.*
Wallace Woolfenden

Right: Aspen groves glowing in fall color along cool, moist slopes near Upper Summer Meadows in the eastern Sierra Nevada.
Rollie Rodriguez

Below: Long-eared owl (Asio otus) *in an aspen grove.*
John Sterling

Glass Mountains support old-growth, if sparser, forests of Jeffrey pine and lodgepole pine forests.

In the interior ranges south of the Glass Mountains and the Tablelands, these montane conifers are rare to absent. White fir and western white pine are absent, while Jeffrey pine, lodgepole pine, and western juniper grow only at single to a few locations each, and these in the White Mountains. Their presence there is likely relictual from past wetter, cooler climatic regimes that supported more widespread distribution of these conifers. At present their occurrence in these arid ranges is limited to alluvial and glacially derived soils where moisture is adequate.

Cool slopes with high water tables and rocky soils support extensive aspen groves along the eastern Sierra Nevada as well as in northern ranges of the California Great Basin. Where conditions are windy and exposed, these can take shrubby forms, growing as dense and stunted thickets that blaze the slopes red, orange, and yellow in autumn. Aspen groves also occur in riparian and meadow-fringe contexts, where trees are upright and have large, straight boles. The capacity of aspen to root-sprout and spread clonally is legendary, but recent genetic research suggests that seedling establishment is also common and aspen clones can be genetically diverse. Aspen groves provide important hiding cover as deer fawning habitat, and many other woodland creatures seek shelter in aspen groves.

Across the western U. S., concern for aspen decline has raised questions about its ecology, pathologies, and conservation, invoking many lines of research. Some of the change in aspen demographics is likely natural: populations are short-lived and dying, even-aged patches are readily apparent on the landscape. Aspen are also sensitive to changes in water levels, and climate dynamics tend to mobilize boundaries of groves readily. Sudden aspen decline (SAD) is of special concern, and has been linked to severe drought and specific environmental contexts (e.g., low, flat aspects) that exacerbate hot and dry conditions. SAD has been observed only in the last decade, and then mostly in the Rocky Mountains, but conservationists are watching for signs of it in California. Excess livestock grazing of young sprouts also inhibits persistence of aspen stands and is a serious conservation issue.

Mountain mahogany shrublands,

a dry, upland counterpart to aspen, extend across low-to-mid elevation slopes of the California Great Basin. Two alliances form these dense shrublands in our region. Curl leaf mountain mahogany (*Cercocarpus ledifolius*) scrub is widespread at cooler middle elevations of the eastern Sierra Nevada and in other northern ranges extending south to the White, Inyo, and Panamint mountains, where they occur on and off carbonate soils. Shrublands have expanded significantly in recent decades, creating fire risk and also serving as cover for mountain lion (*Puma concolor*), mule deer (*Odocoileus hemionus*), and Sierra Nevada bighorn sheep (*Ovis canadensis sierrae*). Conservation plans for recovery of this

Top: Evergreen curl leaf mountain mahogany forms dense stands of large shrubs or small trees along dry slopes of California Great Basin ranges, as here in Lundy Canyon of the eastern Sierra Nevada.

Right: Shrubby aspen stands provide hiding cover for hares, rabbits, young of mule deer, and bighorn sheep high in Parker Canyon.
CONNIE MILLAR PHOTOS

Above: *Lower slopes with curl leaf mountain mahogany scrub are found commonly along the eastside of the Sierra Nevada, as in Big Pine Canyon.*
Julie M. Evens

Below: *Singleleaf pinyon forms extensive woodlands throughout the low elevations of Great Basin mountains, often in pure stands, as in the southern White Mountains looking west to the Sierra Nevada.*
Connie Millar

endangered bighorn sheep include burning stands of mahogany along eastern Sierra Nevada canyons.

Small leaf mountain mahogany (*C. intricatus*) scrub is found on warmer, drier, and generally lower sites. This alliance is far less abundant in the California Great Basin than curl leaf mahogany scrub, and restricted to carbonate substrates in the southeast ranges. When in bloom and backlit by the sun, both species display a shimmering lacey adornment, the delightful effect of hundreds of plume-like styles protruding from the many small flowers.

The lowest elevation conifer forests in the California Great Basin are dominated by singleleaf pinyon (*Pinus monophylla*) woodlands and, with less representation in this region, Utah juniper (*Juniperus osteosperma*) woodlands. Singleleaf pinyon and Utah juniper might be thought of as tree counterparts of sagebrush for their ubiquity across the Great Basin. Short-lived, fire-prone, and prized by native people and wildlife for large, wingless, and protein-rich seeds, singleleaf pinyon are currently considered by some range managers and ranchers as weedy and disruptive of native shrublands. Conditions of the late twentieth century, including climate change, invasive species, and historic land uses, have created a golden opportunity for singleleaf pinyon and Utah juniper establishment and woodland expansion. Along historic range margins, the forest type is rapidly expanding, and infill is

common. Concern for displacement of sagebrush steppe—the prime habitat of greater sage-grouse (*Centrocercus urophasianus*)—directly or by heightening fire risk has become a contentious conservation topic. Aggressive forest management actions are attempting to limit spread and reduce the area of these woodlands throughout the Great Basin. Pinyon-juniper woodlands grade most commonly to sagebrush and other shrub communities at their low-elevation ecotone, creating distinctive lower treelines on many California Great Basin ranges.

The extensive and rather low-diversity singleleaf pinyon and Utah juniper woodlands are sometimes displaced by scattered stands of montane conifers growing far lower than would be expected from their typical range limits. These populations occur on isolated patches of hydrothermally-altered andesitic soils. These exceptional habitats have chemical deficiencies strong enough to exclude many shrubs and herbs, yet montane conifers are able to establish secure if sparse footholds. The woodlands occur as pure or mixed stands of limber pine, Jeffrey pine, and lodgepole pine. The red-to-yellow-to-greenish soils are highly erosive, and reveal severe chemical alterations. Nutrient sterility and toxicities exclude most understory species, making these sites appear even more dramatic. To California botanists these patches might be reminiscent of other pine/cypress types occurring on isolated exposures of sterile ultramafic or sandstone derived soils in the western regions. In the California Great Basin, altered soils and exceptional pine stands occur in the northern ranges. They can be found on low slopes of the Sweetwater Mountains and Bodie Hills, where volcanism was extensive and hydrothermal activity common.

Joshua tree (*Yucca brevifolia*) woodlands, widespread in southeastern California and indicator of the lower tree limit in Mojave Desert mountains, are scattered in the California Great Basin in the Coso, Cottonwood, and Inyo Mountains. There they mix with stands of big sagebrush (*Artemisia tridentata*), shadscale (*Atriplex confertifolia*), and black brush (*Coleogyne ramosissima*). Long-

Above: Pinyon and juniper woodlands expanding into sagebrush scrub.
CONNIE MILLAR

Below: Greater sage-grouse male forages near Crowley Lake.
JOHN STERLING

lived and fast growing, Joshua trees are conspicuous, picturesque, but relatively short-lived, with individuals persisting only 15–200 years. When stands are dense, mature trees scatter across the landscape rather than growing in clusters, which suggests that Joshua trees compete for water taken up by their dense networks of fibrous roots and stored

Above: Joshua tree woodlands, here in the Inyo Mountains, are scattered in disjunct parts of the California southern Great Basin. They can extend to mid-slope elevations in areas that do not get extensive sub-freezing temperatures. Julie M. Evens

Left: Scattered patches of hydrothermally altered soils in the western Great Basin can support montane conifers at elevations lower than typical ranges—Jeffrey pines grow with Utah juniper and singleleaf pinyon in the Bodie Hills. Connie Millar

between rains in the plants. Curiously, dispersal of seeds was at one time provided by the now-extinct Shasta ground sloth (*Nothrotheriops shastensis*), suggesting that adaptive range shifts under present changing climates may be difficult now that the species lacks its dispersal partner.

Montane Shrublands

While forest and woodlands are keystone habitats in California Great Basin mid-montane ecosystems, upland shrublands are extensive and diverse across lower slopes and basins, especially in the more arid south. Most common are the various sagebrush communities, and best known among them is big sagebrush shrubland alliance, which blankets plains, valley bottoms, alluvial fans, ridge slopes, and openings in forests and woodlands. Big sagebrush dominates where soils are sandy, well-drained, non-saline, and deep. The bouquet of sagebrush foliage is beloved widely as the incense of the steppe, and sage is prized by native people for curative powers. Big sagebrush shrubland is critical habitat to many signature Great Basin animals, of particular concern for greater sage-grouse and pygmy rabbit (*Brachylagus idahoensis*), both of which depend on old-growth sage stands. Many subspecies of big sagebrush exist whose identifications challenge the most determined botanist. Bitter brush (*Purshia tridentata*) often is a companion in big sagebrush communities, and, where common, bitter brush forms communities of its own, a situation typical also of rubber rabbitbrush (*Ericameria nauseosa*) and pink-flowering desert peach (*Prunus andersonii*). Mormon tea (*Ephedra viridis*), with its striking yellow-green stems, also commonly comingles with big sagebrush, as does roundleaf snowberry (*Symphoricarpos rotundifolius*), whose white berries

Top: Extensive shrublands of big sagebrush, the iconic plant of the Great Basin, form sagebrush seas from the valley basins below the singleleaf pinyon woodland zone to upland high plateaus; along County Line Hill in the White Mountains.

Above: Fire is an important component to keeping native shrublands healthy, although cheatgrass invasion is drastically altering natural fire regimes in the region.
Connie Millar photos

Above: Rubber rabbitbrush and bitter brush often mix with big sagebrush to form textured and colorful steppe, as along Mono Lake.
Connie Millar

*Below: Silver sagebrush (*Artemisia cana*) with Douglas' sedge (*Carex douglasii*) in a moist meadow below Convict Lake.*
Julie M. Evens

provide important forage for shrubland birds, but are poisonous to humans. Mixed in mesic swales and watered slopes of the northern uplands are montane chaparral communities of tobacco brush (*Ceanothus velutinus*), and mountain whitethorn (*C. cordulatus*), such as in the Glass and Sweetwater Mountains.

Montane Herbaceous Communities

Forests and shrublands of the montane California Great Basin zones are broken by many openings with herbaceous alliances. Often these are highly diverse wet to dry meadows, and species mixes depend on soils, elevation, and degree of moisture saturation. Sedge meadows abound in seasonally or permanently saturated soils of wet meadows, lakeshores, and creeksides at middle to subalpine elevations; these include Sierra alpine sedge (*Carex scopulorum*) turf in Sierra Nevada, Sweetwater Mountains, and Carson/Walker drainages. Adjacent to willows (e.g., *Salix exigua, S. lutea*), water sedge and lakeshore sedge (*Carex [aquatilis, lenticularis]*) meadows are common at many elevations of the eastern Sierra Nevada, Glass Mountains, and Sweetwater Mountains. Other wetland sedge communities are common at middle elevations in the northern California Great Basin, including Nebraska sedge (*Carex nebrascensis*) meadows and beaked sedge meadows around fens, and in seeps and swales of the Truckee, Carson, and Walker drainages and the Sweetwater Mountains.

Mimicking fields of corn to which the common name attests, tall-growing white corn lily (*Veratrum californicum*) unfurls long, corrugated, and deep-green leaves with showy clusters of white flowers tightly packed along branched floral stems. The white corn lily patches alliance occurs on wet soils

near streams, wet meadows, and forest borders. White corn lily is often associated with livestock grazing, given that high levels of alkaloids in its foliage are toxic to large mammals, as a result they are not grazed. An extract from white corn lily, cyclopamine, is being tested as an anti-tumor drug for humans.

Occupying diverse wetland sites, including wet meadows, fens, marshes, and sloughs, Baltic and Mexican rush (*Juncus [arcticus* ssp. *littoralis, mexicanus*]) marshes range from lowlands to high mountains with circumboreal distribution. Plants of Baltic rush lack leaf blades and grow up to 1 m tall. In the California Great Basin, rush marshes occur in Bridgeport Valley, and Fish Slough. Pale spike rush (*Eleocharis macrostachya*) is a matt-forming rush, with tall stems and long floral spikes. Stands are common in sloughs and slow-moving waters near Benton and Bridgeport, for instance, and in seeps and springs of the Inyo Mountains. Similarly, American bulrush (*Schoenoplectus americanus*), and California bulrush (*S. californicus*) marshes are common in many parts of California, and grow, for instance, in Fish Slough, Benton region, and springs of the Argus Range in the California Great Basin.

At mid-montane elevations of the eastern Sierra Nevada, grass communities are common. Widely distributed meadow barley (*Hordeum brachyantherum*) patches, for instance, form moist meadows, swales, and associate with wetland woody vegetation. Broad ecological tolerance gives this grassland frequent appearance in wet upland habitats of the California Great Basin. Also common and widespread across many elevations in California, curly blue grass (*Poa secunda*) grasslands extend along meadow edges and California Great Basin habitats in situations where water tables are shallow.

Above: Glass Creek Meadow in the Owens River Headwaters Wilderness has a mosaic of soil types, from saturated to dry, which are colonized respectively by sedges, grasses, and many flowering herbs.
ROLLIE RODRIGUEZ

Below: White corn lilies spring up along streams and in wet meadows of the Great Basin.
CONNIE MILLAR

Top: Meadows are common in wetlands of the region, often with diverse sedge species, as in Swauger Canyon, Sweetwater Mountains.
Connie Millar

*Above: Yellow-headed blackbird (*Xanthocephalus xanthocephalus*) at Mono Lake.*
John Sterling

Left: Baltic rush, American bulrush, and California bulrush grow in sloughs and still waters of the low basins from Bridgeport through to Death Valley.
Wallace Woolfenden

Riparian Communities

In the cooler, more mesic northern mountains of the California Great Basin, riparian communities are mostly dominated by angiosperm trees, including, as noted above for higher elevations, aspen. Well-watered upland riparian communities are dominated by black and narrowleaf cottonwoods (*Populus trichocarpa* and *P. angustifolia*). They grow with an assortment of willows, most often sandbar willow (*Salix exigua*) and arroyo willow (*S. lasiolepis*), as well as Sierran willow (*S. eastwoodiae*) in the Carson and Walker drainages, and yellow willow (*S. lutea*) and greenleaf willow (*S. lucida* ssp. *caudata*) in the Sweetwater Mountains. Bitter cherry (*Prunus emarginata*), with its red, shiny bark and delicate white flowers, dots the edges of riparian forests.

In more arid mountains, from the Glass Mountains southward, riparian communities are commonly dominated by dense corridors of water birch (*Betula occidentalis*) thickets, whose lime-green, silver-dollar-sized leaves shimmer in the clear sunlight, illuminating the glowing orange bark of the birch stems. Buffalo berry (*Shepherdia argentea*), with leathery blue-grey leaves, is a welcome indicator of high water table in riparian, spring, or wet locations such as in the Bridgeport and Mono Basins, and scattered locations along the eastern escarpment south to about Big Pine Creek. Black willow (*Salix gooddingii*) thickets replace riparian shrubs along the upper Owens Valley and scattered locations farther south. Prior to water extraction, black willows once formed extensive riparian gallery forests along the Owens River.

Above: Narrow riparian corridors along perennial streams are almost always dominated by willow species.
Connie Millar

Below: In a water birch thicket, trees have characteristic orange-hued bark and lime-green leaves in summer.
Julie M. Evens

With newly restored water supplies, new young stands are growing in the lower reaches of the river north of Owens Lake. Along a number of low canyons in the eastern Sierra Nevada from lower Rock Creek southward, ponderosa pine (*Pinus ponderosa*) forms gallery forests that extend into Owens Valley. California black oak (*Quercus kelloggii*) often associates with this forest, giving the visitor a feeling of being in a western Sierra Nevada forest, but with an eastern California escarpment backdrop.

Above: Riparian vegetation in the wetter areas of the eastern Sierra Nevada, such as lower McGee Canyon, often contain black cottonwood as well as water birch. Together with moist lower slopes of aspen, they form a palette of gold in autumn. John O. Sawyer

Left: Creosote bush and white bursage in the Haiwee area, south of Lone Pine. Wallace Woolfenden

Basins, Bajadas, and Playas

In lower, drier, and more southerly reaches of the California Great Basin, sagebrush steppe is replaced by or intermingles with black sagebrush (*Artemisia nova*) scrub mixed with shadscale and other salt bush (*Atriplex* spp.). These extend broadly across low slopes, rocky hills, and bajadas, and along basin bottoms and playa edges. The wide-ranging community crosses diverse soils, including carbonates, sand, clay, and even desert pavement, occupying a spectrum of successional roles, from seral to climax. Occasionally intermingling with salt bush are scrublands dominated by spiny hop sage (*Grayia spinosa*). Like shadscale, spiny hop sage grows in diverse but warm, dry locations, ranging across mountain slopes to basin bottoms, preferring deep alluvial soils that vary from alkaline to calcareous, to sandy. Large stands of desert holly (*Atriplex hymenelytra*) scrub occur on cinder fields and other volcanic and alkaline soils east of Owens Lake and widely through the southeastern parts of the California Great Basin.

Another shrub alliance of dry, lower slopes and upper bajadas is black brush scrub, which is common in the upper Owens drainage north of Bishop, and throughout the southern ranges of the California Great Basin. Scattered stands of Nevada joint fir (*Ephedra nevadensis*) scrub occur in the Alabama Hills above Independence and the southeastern ranges. This clonal species spreads readily and individuals live as long as 100 years. Nevada joint fir scrub stands have extremely high shrub richness, often with as many as 35 associated species within the community.

The dominant shrubland type of American low deserts, creosote bush (*Larrea tridentata*) scrub reaches from the Mojave province into the California Great Basin in low basins such as Eureka Valley and Saline Valley, and finds its northernmost extent just south of Bishop. White bursage (*Ambrosia dumosa*) is a common associate and also dominates the white bursage scrub alliance along old washes and river terraces of dry low-elevation environments, extending down fans, bajadas, and stabilized sand dunes.

In the far eastern section, white bursage scrub occurs on limestone outcrops, but usually these are on sandy or clay-rich soils. In disturbed areas adjacent to creosote bush scrub, especially those heavily grazed, needleleaf rabbitbrush (*Ericameria teretifolia*) scrub spreads from acidic to saline soils. White-flowered rabbitbrush (*Ericameria albida*) is perhaps one of the most restricted of the rabbitbrush clan, only forming stands around the edges of alkaline meadows and springs as at Fish Slough and Grapevine Springs.

Top: Spiny hop sage scrub with spring-blooming annual plants such as pincushion flower (Chaenactis fremontii) *along an intermontane valley and slope of the Last Chance Range, Death Valley National Park.*
JULIE M. EVENS

Above: Shadscale shrub replaces sagebrush steppe in lower drier basins and slopes of the southern Great Basin in California, here in Titus Canyon in the Grapevine Mountains of Death Valley National Park.
WALLACE WOOLFENDEN

*Above: Nevada joint fir, spiny hop sage, and Anderson's boxthorn (*Lycium andersonii*) intermix in diverse shrublands above Eureka Valley in the Saline Range.*
Julie M. Evens

Opposite, top: The east slopes of the Sweetwater Mountains overlook singleleaf pinyon woodlands that are recruiting into sagebrush scrub. Scattered Jeffrey pines in the middle slope dot this eastward view toward the Pine Grove Mountains and Wassuk Range in Nevada.
Connie Millar

Opposite, bottom: Non-native grasses including cheatgrass quickly invade areas after fire that previously contained lush stands of bitter brush, sagebrush, and singleleaf pinyon and other native vegetation in the eastern Sierra Nevada along Highway 89 below Monitor Pass, Mono County.
Julie M. Evens

Conservation and Restoration

Primary among conservation challenges at low-to-mid-elevations throughout the California Great Basin is the expanding tide of invasive cheatgrass and red brome (*Bromus tectorum* and *B. rubens*, respectively). In their native dry, arid-Mediterranean environments, these annual grasses evolved in the presence of grazing mammals and are highly adapted to thrive where grazing pressure is high and fire frequent. Cheatgrass has infiltrated ecosystems of the Great Basin widely over the past century, displacing native herbs and shrubs, altering natural fire regimes, and vastly transforming ecological structure and function. The main culprit in the California Great Basin is cheatgrass, which sets up a highly successful feedback cycle by converting natural fire regimes (responsible for maintaining ecosystem health in native scrub and pine woodlands) into more frequent and flashy unnatural regimes that further promote spread of cheatgrass at the expense of natives. Cheatgrass emerges, blooms, and dries early in the warm season, creating continuous fuels in otherwise discontinuous vegetation. Fires spread rapidly, eliminating native herbs, shrubs, and pines, and enhancing conditions for further cheatgrass spread. In Nevada, cheatgrass and red brome have converted vast slopes and basins of native plant communities into depleted monocultures. Control efforts in the wake of these invasions have been mostly futile. The primary management goals are to minimize entry of cheatgrass and red brome into virgin habitats, or to decelerate the rate of spread of exotics.

Another issue on center stage is spread of two native tree species, singleleaf pinyon and Utah juniper, into sagebrush steppe, which some conservationists and managers interpret as unnatural behavior catalyzed by improper land-use practices.

As mentioned before, these tree species have accelerated in their capacity to recruit, establish, and expand within the past 150 years. The resulting fragmentation and loss of intact sagebrush communities have been considered partly to blame for declining populations of native vertebrates that depend on the shrublands for habitat. In particular, greater sage-grouse is a candidate endangered species under federal law, promoting widespread management efforts to eradicate pine and juniper in areas where they are spreading. Complex interactions uncertainly related to climate change, modified fire regimes, expansion of non-native invasives, and altered hydrology have motivated a widespread yet controversial effort to remove these species where they are expanding.

Constance I. Millar

CHAPTER SIX

Sierra Nevada and Southern Cascades

The Sierra Nevada, a massive mountain chain some 600 km long and 110 km wide and totaling 63,000 square kilometers, covers 42% of the state's entire area. The north to south extent is also impressive, from 35°30′ N latitude at the southern extreme to 40°30′ at the northern extreme. A large part of the Sierran region is characterized by mixed conifer forest, one of California's five largest major habitat types. Some of the state's most productive forests and the nation's most beautiful vistas—Yosemite and Sequoia–Kings Canyon National Parks along with many state parks—are located in the Sierra Nevada and in particular are situated in or near mixed conifer forest. Many first-time visitors to the state think of these mountains and forests as being iconic for California in its entirety, but equally extensive and attractive landscapes exist in lower elevation foothills, valley grasslands, and more arid desert scrub.

This chapter describes the ecology and species composition of the major Sierra Nevada and Cascadian landscapes and alliances, leaving many others not so widely represented to be named but not described at the scale of detail reserved

Above: Mixed conifer forest characterizes the landscape near Bridalveil Falls.

Opposite: Overlooking Yosemite Valley and mixed conifer forest from Glacier Point in Yosemite National Park.
JEFF BISBEE PHOTOS

Top: Open Jeffrey pine forest on a colder and drier east-facing slope exists where lower annual precipitation (80 cm, snowpacks less than 1 m deep) falls, Tahoe National Forest.

Above: More closed red fir forest on a west-facing slope exists in the upper montane zone where annual precipitation is maximum (125 cm, with snowpacks up to 4 m deep), Tahoe National Forest.

JULIE M. EVENS PHOTOS

for the most common types. The most intuitive way to describe vegetation in the Sierra Nevada and Cascades is by elevation belts or zones. The four commonly accepted zones, from lowest to highest, are lower montane, upper montane, subalpine, and alpine. These zones have distinctively different climates and vegetation types. The most important environmental gradients that accompany shifts in elevation are length of the growing season, annual air temperature, annual snowpack depth and duration, annual precipitation, degree of soil development, and natural fire regime. As one moves to the north or the south, the belts shift down or up 172 m for each degree of latitude difference. This rate of displacement is relatively steep compared to other North American mountain chains (Parker 1994). The zones or belts also differ between west-facing slopes and east-facing slopes. West-facing slopes receive more precipitation and tend to be warmer, while east-facing slopes are drier and colder. Displacement of zones on east-facing slopes is higher by about 200 m relative to west-facing slopes at the same degree of latitude.

Each zone contains many vegetation types, some of which are limited to a single zone and are not encountered in any other zone. Individual plant species are similarly limited in their range of belts, being endemic to, or exhibiting maximum cover in, one belt. Although forests cover less than half the total area in lower montane, upper montane, and subalpine zones, they are often informally used to character-

ize each of those three zones. Lower montane forests include mixed conifer, white fir (*Abies concolor*), and giant sequoia (*Sequoiadendron giganteum*) types; upper montane forests contain red fir (*Abies magnifica*) and lodgepole pine (*Pinus contorta* ssp. *murrayana*) alliances; subalpine forests include open mixed conifer woodlands and mountain hemlock (*Tsuga mertensiana*) forests. Important non-forest vegetation types include montane chaparral, wetland willow scrub, aspen (*Populus tremuloides*) groves, wet meadow, rock outcrops, and open herbland on subalpine and alpine fell field slopes.

The Lower Fringe: Mixed Broadleaf Forests

At the lowest fringe of the montane zone (700–1,200 m), discontinuous patches of mixed broadleaf forests occur; more technically called the canyon live oak (*Quercus chrysolepis*), California black oak (*Q. kelloggii*), and California bay (*Umbellularia californica*) forest alliances.

Tree species form two layers. Overstory trees are a mix of ponderosa pine (*Pinus ponderosa*), incense cedar (*Calocedrus decurrens*), and Douglas fir (*Pseudotsuga menziesii*). The two oaks and California bay create a shorter canopy of trees along with tanoak (*Notholithocarpus densiflorus*), madrone (*Arbutus menziesii*) and interior live oak (*Quercus wislizeni*), with minor cover from winter-deciduous Pacific dogwood (*Cornus nuttallii*) and hazelnut (*Corylus cornuta*). Shrubs and

Opposite, top right: White fir–sugar pine forest in the lower montane zone (1,400 m) in Yosemite National Park. Todd Keeler-Wolf

Top: Lodgepole pine forest in the upper montane zone (2,200 m) below Round Top peak, Alpine County.

Above: Mixed subalpine woodland (mainly whitebark pine and lodgepole pine alliances) at 3,000 m. Jeff Bisbee photos

Top: California black oak forest alliance at the low fringe of the lower montane zone in Placer County.

Above: Fall colors in a post-fire, mixed broadleaf forest with bigleaf maple and Pacific dogwood, along the South Fork American River, Eldorado County.

Jeff Bisbee photos

herbs are low in cover and richness, but the diversity of tree species makes for a large number of alliances and associations.

Very little old-growth remains of these alliances, both in the Sierra Nevada and the Coast Ranges, mainly because of disturbance from extensive past logging. Following disturbance, all broadleaf trees sprout at their base, but conifers must re-invade by seed, making early succession a lengthy process. Once conifers become established, however, Douglas firs have the capacity to rapidly grow up through the hardwood canopy, and they do not have to wait for canopy gaps to form (Hunter and Barbour 2001). The fire regime may vary from stand to stand, but overall fires are of moderate intensity and severity, with low to moderate complexity, and they recur every 30–100 years.

Mixed Conifer Forests

With increasing elevation (1,000–1,800 m) the mixed broadleaf forests give way to mixed conifer forests, characterized by six species: ponderosa pine, Douglas fir, incense cedar, sugar pine (*P. lambertiana*), white fir, and—in roughly 60 stands that are most common in the southern Sierra Nevada—giant sequoia. "Mixed" means that stands vary in dominant species, and the identity of a dominant tree—even the presence of a half-dozen tree species—changes from hectare to hectare. Generally, though, the predominant mixtures are the ponderosa pine–Douglas fir, ponderosa pine–incense cedar, white fir–sugar pine, and white fir–Douglas fir forest alliances. These are structurally complex forests, with four canopies. Below a conifer canopy is a very open sub-canopy of winter-deciduous hardwoods such as black oak and dogwood, then a patchy canopy of evergreen shrubs such as green leaf manzanita (*Arctostaphylos patula*) and shrub tanoak (*Notholithocarpus densiflorus* var. *echinoides*), and finally a ground-surface canopy of shorter shrubs such as pinemat manzanita (*Arctostaphylos nevadensis*), mountain misery (*Chamaebatia foliolosa*), and glory mat (*Ceanothus prostratus*) mixed with scattered perennial grasses and forbs. Such architecture provides habitat for a high diversity of birds and mammals.

These stands have been strongly and negatively affected by a century of fire-suppression management during which all fires—no matter how small or non-threatening to humans and their structures—were extinguished as quickly as possible. The natural fire regime of small, cool-burning surface fires with low intensity and severity, returning every 25 years, was disregarded. As a result, the density of white fir adults has increased five-fold and that of juvenile ponderosa pine has declined almost as sharply. What had been open, sunny, mixed conifer forests is becoming a closed, dense, white fir forest alliance with a fire return interval of 100–600 years. Fortunately, mixed conifer forests with a natural fire regime still in place exist in the Sierra de San Pedro Martir of Baja California at the southern tip of the Peninsular Range. Approximately 60,000 hectares of old-growth forest exists within the protective borders of a national park. The area is

Above: Open canopy of a mixed conifer forest with incense cedar, ponderosa pine and white fir with patches of shrubs in Yosemite National Park.
John Stuart

Left: A young fisher (Pekania pennanti) *near Kings River, southeast of Shaver Lake in Fresno County. Fishers prefer dense, structurally diverse forests at lower elevations.*
Zane Miller

Old-growth mixed conifer forest dominated by Jeffrey pine in the Peninsular Range of Baja California. This ecosystem has a species composition, geological setting, and climate very similar to that in the Sierra Nevada. Michael E. Kauffmann

so remote that it has never experienced fire suppression management, timber harvest, or over-grazing by domestic livestock. The forests there are like a window through time, providing a landscape-scale glimpse of what Sierra Nevada forests may once have been a century ago. These modern Baja forests bring to mind John Muir's description of nineteenth-century Sierran forests as being open enough for horses and wagons to pass easily and the structure of the forest making it difficult for a slow-moving surface fire to grow into a catastrophic crown fire.

In well-defined patches—only on west-facing slopes and usually in bowl-shaped areas with a shallow water table—are stands of the giant sequoia forest alliance. These groves extend from Placer County in the north to Tulare County in the south, increasing in size and number from north to south. The largest groves contain more than 20,000 trees and the smallest contain fewer than a dozen trees. Adult trees are the most massive living things that have ever existed on the planet: 80 m tall, bole diameters up to 5 m across, and biomass of approximately 6,000 metric tons. The trees are also long-lived, with as many as 3,266 annual rings.

The species has apparently evolved in a fire-type climate, because the heat of surface fires is necessary to open the small, numerous cones. Once the layers of green, living cells have been killed by the heat, water tension no longer keeps the cone scales closed. The open cones remain attached to the parent tree and the tiny seeds are scattered to the wind. Giant sequoia is, then, a serotinous species (meaning late to open). Surface fire also removes litter, allowing seeds to fall directly on bare mineral soil. The root of a germinating seed can then immediately grow into mineral soil, which retains moisture far longer than litter does. Seedling establishment is not successful when germination occurs on litter.

Another kind of serotinous species in the lower montane belt is knobcone pine (*Pinus attenuata*), which thoroughly dominates the knobcone pine forest alliance. Knobcone pine is an obligate fire-type tree, keeping unopened cones firmly attached to canopy branches for decades until a crown fire kills all the adults and induces cone scales to open when high temperatures melt a cement-like resin that permeates the cones. The architecture of the trees and the forest favor crown fires: trees are dense and spindly, have thin bark, and retain numerous side branches all the way down to the ground. A great deal of dead litter builds up, fueling fire that carries it into the canopy. Once the fire has died, the soil surface is rich in ash, support for a dense cohort of newly germinated pine seedlings, which have no other living plants with which to compete. The site is "relayed" to the next generation. One generation is killed, and the next is established under the best possible habitat conditions. The fire return interval is 20–50 years, and no individual older than 90 years has ever been documented. The site repeatedly experiences fire, seed rain, massive pine germination and establishment, the poor growth of a very dense forest, and a relaying of the site from one generation to the next.

As elevation rises to the transition ecotone between lower and upper montane zones, mixed conifer stands contain increasingly more and more red fir and shift to stands of the red fir–white fir forest, with much lower cover by herbs and shrubs and a lower diversity of tree species in the overstory.

Upper Montane Forests

At approximately 2,000 m elevation, an enormous transformation of vegeta-

Top: A knobcone pine stand above the South Fork of the Merced River.
Julie M. Evens

Right: Giant sequoias in the Mariposa Grove in Yosemite National Park.
Jeff Bisbee

Top: Wetter habitats support lodgepole pine forest and beaked sedge and blister sedge meadows alliance along Horse Creek in Mono County.
Jeff Bisbee

Above: The upper montane zone with red fir forest alliance accumulates more snow than any ecosystem in California. Note the yellow wolf lichen, Letharia vulpina, *which extends down the trunk until it reaches average snowpack depth, below which it cannot maintain itself.*
Robert Wick

tion occurs. The architecture of forest stands drops from four canopy layers to one, and the species composition of tree, shrub, and herb layers largely differs from that in the lower montane. The fraction of annual precipitation that falls as snow triples, the growing season shrinks in half (from 7–9 months to 4–6 months), the fire return interval doubles to 50+ years, and the vegetation is no longer positively invigorated by surface fire as it was in the lower montane.

The most common mesic forest vegetation of the upper montane zone is that of the red fir forest alliance. Wetter habitats support stands in the lodgepole pine forest alliance; rockier and drier sites are occupied by stands in the Jeffrey pine (*Pinus jeffreyi*) forest and western white pine (*P. monticola*) forest alliances and mountain juniper (*Juniperus grandis*) woodland alliance. Red

fir can be present in these alliances and may even share dominance with the pines.

Red fir stands are one-layered: an overstory of conifer trees, comprising more than 40% cover. There are no understory hardwoods and there is very light cover from shrubs and mainly perennial herbs. Red fir is one of the largest species of fir in the world, able to attain diameters of 2 m, heights of 47 m, and ages of 600 years. Although surface fires do scar many mature trees, there is no evidence that red fir responds positively to fire. Furthermore, red fir individuals rarely show repeated scars, indicating that they may not survive multiple fires in their lifetime.

This zone has the deepest, longest-lasting snow packs (up to 4 m deep and present for 200 days) of any place along the west coast of North America. There is some evidence to show that winter conditions drive the change in species in the relatively narrow ecotone between lower and upper montane zones (Barbour et al. 1991).

Recovery from logging of red fir forest is also different from succession in lower montane forests. Lower montane pines germinate and grow best in full sun, consequently clear-cutting (all trees taken) or select cutting (20–60% of trees taken) provide those open conditions. Optimum conditions for regeneration of red fir are deep shade (at least for the first dozen or so years of sapling life, after which optimum conditions are full sun). Consequently, strip cutting (all trees cut in a 10 m-wide strip oriented north-south so that as the sun moves from east to west, the narrow strip is covered by shadows most of the day) is the best management option for red fir stands.

Drier montane and subalpine bedrock outcrops support mountain juniper woodland alliance.
Julie M. Evens

Montane Chaparral

Montane chaparral replaces forests where environmental conditions (such as soil depth, slope aspect, soil moisture storage, disturbance history, and geological substrate) are too xeric or unfavorable for forest vegetation. Montane chaparral differs from low elevation chaparral in two ways. Unlike the climax-like permanence of foothill chaparral, montane chaparral can exist as a relatively short-term successional stage following fire and deforestation. It can also be stable on dry, skeletal soils, on south-facing slopes, and on low-nutrient substrates. Second, montane chaparral is floristically different at both the generic and species levels. For example, the genera *Adenostoma* and *Rhus* are absent in montane chaparral, and species of *Ceanothus* and *Arctostaphylos* differ among the two types of chaparral.

Several alliances have been recognized in the Sierra Nevada, depending on which shrub species dominates, such as the green leaf manzanita chaparral, mountain whitethorn (*Ceanothus cordulatus*) chaparral, deer brush (*C. integerrimus*) chaparral, tobacco brush (*C. velutinus*) chaparral, bush chinquapin (*Chrysolepis sempervirens*) chaparral, Brewer oak (*Quercus garryana* var. *fruticosa*) scrub, and huckleberry oak (*Quercus vacciniifolia*) chaparral alliances. These alliances are common in both upper

Top: Green leaf manzanita chaparral with huckleberry oak and bush chinquapin on rocky soils, next to white fir and Jeffrey pine forest, along Lake Tahoe near the California–Nevada border. Jeff Bisbee

Middle: Post-burn deer brush chaparral with emerging ponderosa pine, California black oak, and other trees near Wrights Lake in the Eldorado National Forest.

Left: Montane chaparral dominated by tobacco brush post burn in the Plumas National Forest. Julie M. Evens photos

and lower montane zones.

Prior to the twentieth-century "no burn" forest policy, montane chaparral was historically more widespread. Ecologically, fire releases soil nutrients and creates openings, allowing for a diverse matrix of shrub, herbaceous, and tree growth. Within a few weeks after fire, some species of shrubs resprout from underground stems while others reproduce from seed based on certain adaptive triggers for germination, such as heat from the fire or chemicals released from burned wood. Also, fire-following wildflowers germinate, bloom, and set seed in incredible displays. Nitrogen-fixing chaparral shrubs (*Ceanothus* spp.), white alder (*Alnus rhombifolia*), and lupines (*Lupinus* spp.) can further ameliorate the soil. Early seral forests establish in complex arrangement with shrubs, herbs, and tree snags within 20 years or longer, and they provide habitat for many other species including insectivorous birds, cavity-nesting birds, rodents, and various bugs that are attracted to the flowers, foliage, seeds, logs, snags, and other features of this post-burn landscape (Longstreth 2014).

Wetlands and Meadows

Montane wetlands include riparian thickets, groves of hardwood trees, wet to dry meadows, fens, and seeps. In general many of these wetland types and their species occur in lower montane and subalpine zones as well as in the upper montane. Wet meadow borders are occupied by red osier (*Cornus sericea*) thickets, aspen groves, mountain alder (*Alnus incana*) thickets, and several other alliances dominated by willows such as Sierran willow (*Salix eastwoodiae*), Geyer willow (*S. geyeriana*), Lemmon's willow (*S. lemmonii*), Sierra gray willow (*S. orestera*), and Jepson willow (*S. jepsonii*).

High elevation (upper montane, subalpine, alpine) Sierra Nevada meadows

Top: Mountain whitethorn, Placer County.
Jeff Bisbee

Below: A Sierra Nevada yellow-legged frog (Rana sierrae) *scans the landscape in Yosemite National Park.*
Devin Edmonds, USGS

contain more than 30 herbaceous plant alliances, and any given meadow may contain half a dozen of those alliances, all differing in the length of inundation time and depth to the water table. Common sedge-dominated meadow vegetation includes water sedge and lakeshore sedge (*Carex [aquatilis, lenticularis]*) meadows, Sierra alpine sedge (*C. scopulorum*) turf, and beaked sedge and blister sedge (*C. [utriculata, vesicaria]*) meadows, among many others. Forb-dominated vegetation includes western cowbane (*Oxypolis occidentalis*) meadows, western false asphodel–California bog asphodel (*Triantha occidentalis–Narthecium californicum*) fens, white corn lily (*Veratrum californicum*), and herb-rich (*Senecio triangularis*) meadow alliances. Various other alliances occur with these forb-dominated types along with many mosses in highly saturated, peat-rich fens, including bog blueberry (*Vaccinium uliginosum*) wet meadows, tufted hair grass (*Deschampsia cespitosa*) meadows, few-flowered spike rush (*Eleocharis quinqueflora*) and Sierra rush (*Juncus nevadensis*) marshes, Pacific bog sedge (*Kobresia myosuroides*) meadows, woodland sedge (*Carex luzulina*) fens, short-beaked sedge (*Carex simulata*) meadows, and shore sedge (*Carex limosa*) fens. The perennial water supply provides refugia for plant and animal species that persist only in fens—many of these species have their main ranges of distribution far to the north in Alaska and Canada and their

Above: Aspen grove along the border of a wet meadow in Hope Valley in Alpine County.
Bob Lyons

Left: Lemmon's willow thicket wetland, Ebbetts Pass, Calaveras County.
Julie M. Evens

southernmost range in California.

Dry meadow vegetation occupies sites where soils become dry earlier in summer than those of wet meadows. Soils are coarse enough to have a relatively deep water table, too deep for plant roots to ever reach saturated soil layers. Many different herbs grow in such meadows, including species in the shorthair sedge (*Carex filifolia*) turf, shorthair reed grass (*Calamagrostis muiriana*) meadow,

Top left: Beaked sedge dominates a meadow near Utica Reservoir.
Todd Keeler-Wolf

*Top right: Shorthair reed grass with dwarf bilberry (*Vaccinium cespitosum*) at Yosemite National Park.*

Right: Tufted hair grass meadows alliance is characteristic of wet meadow edges.
Julie M. Evens photos

black alpine sedge (*Carex nigricans*) sod, and pussypaws–groundsmoke (*Cistanthe* [*umbellata*]–*Gayophytum* [*diffusum*]) herbaceous alliances. Mule ears (*Wyethia mollis*) can dominate large areas of very dry meadows, accompanied by very few individuals of other species, possibly as a result of allelopathic substances released by mule ear plants. Dry rock outcrops also have their own specialized set of subshrubs and herbs, including patches of Parry's rush (*Juncus parryi*) and mountain pride (*Penstemon newberryi*).

In some cases, trees can tolerate the edge of wet meadows during dry years, but then they suffer from anoxia (lack of oxygen) in the root zone when average or above-average precipitation returns. Such vegetation leaves zones of dead mature trees (where wet years had little effect until an exceptionally wet year killed adult trees) and strips of dead young trees (where tree establishment occurred too far into the meadow, where flooding is more frequent). The most common forest vegetation that occupies this tension zone between wetlands and uplands is the lodgepole pine forest alliance. Other trees, not so abundant nor as dramatically responsive to wet vs. dry years, are aspen and black cottonwood (*Populus trichocarpa*).

Subalpine

The subalpine zone extends from 2,600 to 3,300 m elevation. Here, the growing season shrinks to three months and annual precipitation drops to 100 cm (while annual precipitation in the upper montane, where winter snow packs are meters deep, averages 150 cm) because

Top: Sierra alpine sedge in a dry meadow on Dana Plateau.
Todd Keeler-Wolf

Left: Grass Lake, south of Lake Tahoe, is the most extensive basin fen in California, and contains a diversity of fen meadow types.
Julie M. Evens

Above: The Mildred Lake area within the Convict Creek drainage supports rare, nutrient-rich fen meadows that include Pacific bog sedge, shortfruit willow (Salix brachycarpa), and alpine meadow-rue (Thalictrum alpinum), along with an incredible diversity of subalpine vegetation such as limber pine and whitebark pine.
Jeff Bisbee

Left: Mountain pride patches are found amongst large granite expanses, including along the trail to Kaweah Gap with Hamilton Lake and Angels Wing in the distance, Sequoia National Park.
Julie M. Evens

Above: Foxtail pine in the subalpine zone, Inyo County.
Jeff Bisbee

Below: "Multi-trunked" whitebark pine with lodgepole pine and mountain hemlock, Alpine County.
Julie M. Evens

most of the water that could be wrung out of storm clouds coming upslope from the northwest has already fallen in the Coast Range or in mid-elevation zones of the Sierra Nevada. Another distinct difference from lower zones is that here there is essentially no fire regime. The low vegetation biomass does not easily carry fire, so the fire return interval is hundreds of years long.

Subalpine trees are usually scattered in open woodlands (20–40% canopy cover) dominated by a handful of conifers: western white pine, whitebark pine (*Pinus albicaulis*), foxtail pine (*P. balfouriana*), limber pine (*P. flexilis*), lodgepole pine, and mountain hemlock (*Tsuga mertensiana*) alliances. As their species ranges overlap, the complex vegetation is often informally called mixed subalpine woodland. These trees grow slowly because of a short, cool growing season, and when mature they are only 10–15 m tall, but they have been twisted into fascinating asymmetrical shapes by strong winds that have buffeted them throughout their long life spans of a millennium or more.

These subalpine woodlands are basically one-storied. Litter does not accumulate, and the cover by herbs and low shrubs is modest. Furthermore, no hardwoods define a second tree canopy layer. Many trees appear to have multiple trunks, especially as elevation rises to treeline. Research shows that these multiple trunks are not from a single tree, but rather represent a cluster of trees that germinated together from a seed cache deposit that was forgotten by a bird or small mammal.

At the elevational limit for tree

growth, trees become shorter and eventually they become low and hedge-like, about 1 meter tall. This peculiar growth form is called "krummholz," a German word meaning crooked wood. The shrubby trees are not genetically dwarfed: they are forced into a semi-prostrate form because in winter any twigs that are exposed above the snow pack are blasted and killed by ice particles and blown along the surface by strong winds. If seeds from krummholz individuals are collected and then germinated in a protected garden downslope, they all produce trees with a dominant, straight, tall single bole. Not every subalpine conifer in California is capable of forming krummholz; whitebark pine is the most common krummholz conifer in the Sierra Nevada.

Alpine

The alpine zone is dominated by perennial herbs and dwarf shrubs. Trees are no longer present, even as krummholz. Precipitation continues its drop from the upper montane, with an average of 30 cm, virtually all of which falls as snow. Snow packs accumulate only in the shelter of trees and rocks. No snow accumulates at all on barren ridges, making for a very arid situation. The growing season is two months long at best, the most mesic sites being just below melting snow packs. Mean annual temperature is only 1°C, and frost can occur any day or night during the growing season.

Some 250 plant species occur exclusively in the alpine, while another 350 species extend downslope from the alpine into subalpine and upper montane

Krummholz growth form of whitebark pine on the Dana Plateau. The "trees" are shorter than 1 m.
Todd Keeler-Wolf

Top: In the alpine zone there are no longer trees, and vegetation cover can be sparse on fell-fields and exposed slopes.
Robert Wick

Above: Cushion plants are characteristic of the Sierra Nevada alpine zone, with an above-ground portion typically less than 10 cm tall. The below-ground portion has 2–3 times as much biomass as the stems, leaves, and flowers combined.
Michael G. Barbour

zones. Some of the characteristic plants are small, tightly packed forbs called "cushion plants" that have extensive root systems but only a miniature canopy mix of small leaves and large flowers, no taller than 5–10 cm above the ground. Other common growth forms are grasses and sedges.

Habitats include open, stony scree slopes and fell fields, wet meadows, and rocky ridges. Plant cover can reach 100% in wet meadows, but only 15% in fell fields, and 1% on exposed ridges. Common vegetation types in fell fields include pink saxifrage (*Saxifraga nidifica*) patches, Coville's phlox (*Phlox covillei*) fell-fields, heretic penstemon (*Penstemon heterodoxus*) patches, mountain sorrel (*Oxyria digyna*) patches, alpine fescue (*Festuca brachyphylla*) fell-fields, and California heath-goldenrod–Pacific alpine gold (*Ericameria discoidea–Hulsea algida*) fell-fields. Common vegetation types in meadows include several sedge alliances (*Carex breweri, C. helleri, C. spectabilis, C. straminiformis, C. subnigricans*) and prostrate willow alliances (*Salix nivalis* and *S. petrophila*).

Eastern Sierra

Vegetation on the desert-facing side of the Sierra Nevada for the most part also occurs on the west-facing side. However, the spatial separation of subalpine from upper montane is blurred on the east, possibly because the escarpment is steeper, hence the environmental gradient itself is blurred and less abrupt. There are few differences in alliances between the two aspects until the lower montane zone is reached at 2,600–1,700 m elevation.

The wooded portion of the east-side lower montane zone is a mosaic of mesic stands of Jeffrey pine forests and drier stands of ponderosa pine forests, Washoe pine (*Pinus ponderosa* var. *washoensis*), and singleleaf pinyon (*P. monophylla*) woodlands. Singleleaf pinyon woodlands often occur on volcanic substrates, while Jeffrey pine forests often dominate granitic substrates.

Stands in the Jeffrey pine forest have three canopy layers: an overstory conifer canopy 30 m tall that provides 65% cover from a mix of Jeffrey pine, incense cedar, white fir, and sometimes ponderosa pine; a subcanopy 10–15 m tall of singleleaf pinyon with varying amounts of junipers (in particular *Juniperus osteosperma*); and a layer 1–2 m tall of shrubs, some with Sierran affinities and others with cold desert affinities, altogether providing 20% or more cover. The herb layer is sparse. Below the elevational limit for Jeffrey pine forests, at about 1,600 m, is the zone of pinyon-juniper woodland. At 1,400 m elevation, this woodland gives way to cold desert scrub, characterized by the big sagebrush (*Artemisia tridentata*) shrubland alliance. Here, trees are restricted to the riparian zone and their species diversity is low.

Washoe pine woodlands are scattered along the east-facing slopes in the northern part of the Sierra Nevada: Lassen Volcanic National Park, Bald Mountain in Sierra County, Mt. Rose just northeast of Lake Tahoe, and Last Chance Creek in Plumas County. In the Warner Mountains,

Top: The Jeffrey pine forest alliance is very common in the eastside lower montane zone.

*Right: Alpine willow (*Salix petrophila*) turf on the magnificent fell-field at Dana Plateau.*
TODD KEELER-WOLF PHOTOS

northeast of the Sierra Nevada, Washoe pine is much more widely distributed. Whether Washoe pine is a separate species or a variety of ponderosa pine is debatable (Haller and Vivrette 2011). Its major morphological distinction is the small size of its cones. The upper limit to the lower montane zone is approximately 2,200 m, where Jeffrey pine dominance is replaced by denser mixed red and white fir forests.

The Southern Cascades

In many ways, the vegetation of the Cascade Range in California—from Lassen Volcanic National Park to the Oregon-California border—is geologically and geographically transitional between the granite-dominated Sierra Nevada, the volcanic Cascade Range in Oregon, and the more coastal and geologically varied Klamath Mountains. In the lower montane zone, at 1,300–1,700 m elevation, the important tree species are white fir, ponderosa pine, Jeffrey pine, and Douglas fir; less common trees are incense cedar and sugar pine. Forest vegetation falls mainly in the Douglas fir, ponderosa pine, ponderosa pine–Douglas fir, and white fir–sugar pine forest alliances. Historically, a natural surface fire regime existed, similar to that in the Sierra Nevada, but with additional disturbances from volcanic mudflows over the past 1,500 years. Recovery from mudflows is very slow compared to recovery from surface fires. Hardwoods that continue north from the Sierra Nevada include canyon live oak, California black oak, Pacific dogwood, hazelnut, and bigleaf maple

Top: Pinyon-juniper woodland forms the lower timberline on the eastside of the Sierra Nevada giving way to cold desert scrub as elevation declines.
Jeff Bisbee

Left: Washoe pine near Babbitt Peak.
Todd Keeler-Wolf

(*Acer macrophyllum*). Montane chaparral in this zone is typically dominated by tobacco brush and green leaf manzanita. Fires in the last century have increased the area covered by montane chaparral. Deer brush is uncommon, in contrast to its abundance in the Sierra Nevada.

The upper montane zone, 1,700–2,000 m elevation, is characterized by forests of Shasta red fir (*Abies magnifica* var. *shastensis*), white fir, and Jeffrey pine. Shasta red fir is thought to be a hybrid of red fir and noble fir (*Abies procera*)—the latter being a tree of the Cascades to the north in Oregon and Washington. Barbour and Woodward (1985) showed that Shasta red fir behaves ecologically like red fir. The two have stands with a similar age structure that indicates regeneration does not depend on disturbance such as fire and both are associated with a similar group of shrub and herbaceous species. Stands of all three red firs are included in the red fir forest alliance. Montane chaparral in this zone tends to be dominated by huckleberry oak, green leaf manzanita, and pinemat manzanita. Meadows are restricted to the very wettest locations because the volcanic soils are very porous and well-drained.

Above 2,000 m is a subalpine zone. Tree density and canopy cover, as well as

Above: Many hardwood and conifer species found in the Sierra Nevada region continue north to the southern Cascades region.

Below: Red fir forest with pinemat manzanita in the understory.

Jeff Bisbee photos

Above: Although timberline of the southern Cascades, including Shasta red fir, mountain hemlock, and whitebark pine, is lower than in any other mountain ranges of California, the relatively large alpine zone on Mount Shasta and Lassen Peak has a relatively depauperate flora due to the recent volcanic activity, harsh climate, and the porosity and instability of the substrate.
Jeff Bisbee

*Right: Vegetation of the subalpine zone along Lassen Peak includes bluntlobe lupine (*Lupinus obtusilobus*), Davis' knotweed (*Polygonum davisiae*), and other perennials, unlike the subalpine in the Sierra Nevada, but similar to other openings on volcanoes northward through Oregon.*
Ayzik Solomeshch

Above: View from Brokeoff Mountain to Lassen Peak. This mountain is part of an older and larger volcano called Mount Tehama, which has been eroded away over the past half million years by massive ice age glaciers and volcanic forces. However, the relatively recent origins of the mountains, harsh climate, and rapidly receding ice and snowfields all contribute to the scarcity of alpine vegetation here and on all Cascade volcanoes. Lassen Volcanic National Park, Tehama and Shasta Counties. GINA RADIEVE

Left: The Glass Mountain obsidian flow abruptly abuts the older Cascades' volcanic highlands cloaked with a dense forest of mountain hemlock and western white pine. Medicine Lake Highlands, eastern Siskiyou County. TODD KEELER-WOLF

understory plant cover, are greater here than in the mixed subalpine woodland of the Sierra Nevada. The most abundant forest alliances are mountain hemlock and whitebark pine. Pines play more of a role in the Sierra Nevada than in the Cascadian subalpine. Foxtail and limber pines are absent and western white pine is mainly limited to the upper montane.

Alpine zones in the Sierra Nevada and southern Cascades have many taxa in common, but others reach their northern or southern limits on Lassen Peak (and other peaks in California). For example, purple mountain-heath (*Phyllodoce breweri*) reaches its northern limit and pink mountain-heath (*P. empetriformis*) reaches its southern limit on these peaks. Alpine meadows are restricted to high slopes on volcanoes, at elevations where the growing season is twice as long as in the Sierra Nevada—four months versus two months—no doubt due at least in part to more summer precipitation. However, the alpine flora is more depauperate than the Sierra Nevada due to high snow and ice cover, homogeneous and unstable porous substrate, and isolation of individual mountains.

Conservation and Restoration

Human activities of logging, urbanization, road building, setting animals to graze, altering hydrologic regimes, and changing natural fire regimes have

Top: Alpine meadows are found on high elevation volcanic soils of the southern Cascades, including at Lassen Volcanic National Park with purple mountain-heath.
Julie M. Evens

Left: The 2014 Rim Fire in the Stanislaus National Forest has been the third largest wildfire in California. It spread over 104,000 hectares (over 1,000 square km) with severe burning to forest overstories.
U. S. Department of Agriculture

all strongly affected vegetation change in the Sierra Nevada and the southern Cascades. Some of these changes appear to be ecologically neutral, but others have created landscapes that experience slumping and erosion, or support forests so dense that surface fires can easily ladder up trees to become catastrophic crown fires. Overgrazing of meadows has caused the formation of down-cut streams and stringers that are no longer capable of maintaining the diverse types of meadows that characterized these mountain chains of 150 years ago. Restoring the altered landscapes to their previous more sustainable nature and biotic richness will not be easy or quick.

Conservation in state and national parks and U.S. Forest Service Research Natural Areas has been relatively successful, thanks to the fact that the great majority of forested land in the Sierra Nevada and southern Cascades is protected by public agencies, meaning that new ideas in ecosystem management can be applied uniformly throughout the region, and the bottom-line objective can be sustainability of forests rather than short-term revenue generation. Gradually, the U.S. Forest Service has come to see itself as more of a conservation-restoration agency with less focus on the consumption of natural resources (lumber, clean water and air).

Unfortunately, conservation in California can lead to exploitation elsewhere. California forests satisfy only about 70% of the state's demand for wood. Our forests do not satisfy the demand because our standards for harvest are stringent. Where does the missing 30% come from? Undoubtedly it comes from places where forests are over-harvested and mismanaged in order to satisfy local demand and California's demand. We need to foster conservation of resources at a local and global scale, in order to eliminate pockets where the financial bottom line still trumps conservation.

Michael G. Barbour
John O. Sawyer

The Forest Service conducts thinning in this lodgepole pine forest as a restoration effort in the Big Meadow Creek Watershed, south of Lake Tahoe.
U. S. Forest Service, Pacific Southwest Region

CHAPTER SEVEN

Sierra Nevada Foothills

Top: Rolling foothill slopes include oak and pine woodlands, chaparral, and grasslands in Mariposa County.

Above: Vegetation sorts itself along the different slope exposures and rock types in the foothills.
JULIE M. EVENS PHOTOS

Opposite: The North Fork American River, designated as a Wild and Scenic River, bends through hillsides of mesic chaparral and oak woodlands near Auburn.
ROBERT WICK

The long slender belt of the Sierra Nevada foothills arises east of the Great Valley at 60 m and goes to 1,500 m (or higher) in elevation. The San Emigdio Mountains and southwestern Tehachapi Mountains in Kern County form its southern extent, and the southwestern tip of the Cascades Range in Shasta County forms its northern edge. This distance is about 745 km, about 465 miles. Rain—the primary form of precipitation—varies from around 25 to 100 cm (from south to north) annually in the region. While winters are cool and wet, summers are dry and hot.

As the Sierra Nevada has arisen along major north-trending fault zones, a variety of volcanic and plutonic rocks have erupted and subsequently eroded along its mountainsides. This western side of the range has a long, gentle slope as compared to the eastside because the mountains have lifted up steeply in the east along an enormous, tilted fault block. Ice-age glaciers have subsequently worn down the mountains' sides and valleys, where the foothills have been weathering for more than 100 million years. The hills are primarily cloaked with oak woodlands, grasslands, and chaparral along masses of weathered, exposed bedrock of gabbro, granite, volcanic ash, and basalt.

Softer sedimentary rocks also occur in the foothills, including ocean floor sediments that have historically been dragged into the Sierran trench as North America's continental plate has

Top: Scattered outcrops of serpentinite in the foothills contain an array of herbs along with the hardy wedge leaf ceanothus chaparral and foothill pine.
Julie M. Evens

Above: The Ione sandstone formation supports an assortment of rare shrubs and herbs in the Ione manzanita alliance.
Jeff Bisbee

moved westward. Interesting and ancient deposits of gravels, mudstone, and sandstone occur scattered along the low western front of the Sierra Nevada, while others have been metamorphosed into gray slates and greenstones (Alt and Hyndman 2000). Many of these scattered sedimentary rock formations have unusual plant assemblages. For example, the Ione sandstone formation contains cement-like crusts of yellow iron oxide where endemic and rare assemblages of Ione manzanita (*Arctostaphylos myrtifolia*) and Ione buckwheat (*Eriogonum apricum*) occur along with other chaparral and interior live oak (*Quercus wislizeni*) woodlands.

In contrast, slices of serpentinite rocks occur as dark-green slabs of metamorphosed, ancient oceanic crusts along fault zones within layers of gabbro and basalt dikes. Serpentinite soils contain heavy metals, including magnesium, which are harsh for many plants; however, whiteleaf manzanita (*A. viscida*), wedge leaf ceanothus or buck brush (*Ceanothus cuneatus*), and foothill pine (*Pinus sabiniana*), as well as lovely displays of native wildflowers, can tolerate these soils.

Prevalent Woodland and Forest Types

The quintessential vegetation type of the foothills is blue oak (*Quercus douglasii*) woodland alliance, often underlain by and dappled with grasslands. Whether blue oak is alone or intermixed

with other typical foothill trees, stands cover much of the rolling foothills. Pure blue oak woodlands are usually found in warmer settings, as compared to pure interior live oak woodland or mixed live oak and foothill pine stands that more regularly occur on moister, northerly slopes and draws. Mixtures of blue oak and interior live oak are also common where moisture is intermediate in nature, while valley oak (*Q. lobata*) woodland alliance occurs more scattered along or adjacent to creeks and other moist settings.

Foothill pine typically occurs intermixed with blue oak and interior live oak across most of the foothills. However, pure stands of foothill pine woodland alliance do occur in certain environments—particularly on serpentinite and gabbro substrates. The pines tolerate these nutrient-poor substrates while the oaks cannot. Foothill pine is absent in an 89-km wide gap in the southern foothills (mostly in Tulare County), yet no agreed upon theory currently exists to explain it. Herbs are common understory plants of the more open oak and pine woodlands, and a variety of natives and non-natives occur in this habitat. Other common associates include shrubs of wedge leaf ceanothus, common whiteleaf manzanita (*Arctostaphylos manzanita*), whiteleaf manzanita, toyon (*Heteromeles arbutifolia*),

Top: Blue and interior live oak woodlands cloak the low foothill slopes at McKenzie Table Mountain Preserve.

Above: Foothill pine casts a slender but regular shadow upon rocky knolls of serpentine, in contrast to the blue oak lower down on deeper, more nutrient rich soils.

JULIE M. EVENS PHOTOS

Top: The mounded crowns of California buckeye cling to the edges of this table mountain near the Stanislaus River in Tuolumne County.
Jeanne Taylor

Above: California black oak and canyon live oak forests with ponderosa pine commonly occur in the upper foothills with a lovely variation in color.
Julie M. Evens

hollyleaf redberry (*Rhamnus ilicifolia*), and poison oak (*Toxicodendron diversilobum*).

California buckeye (*Aesculus californica*) woodland alliance also emerges in distinct habitats: moist concave or north-facing slopes and rocky volcanic slopes below tablelands. This broadleaf tree is unique in being a summer-deciduous plant that loses its leaves in the hottest part of the year to overcome moisture stress.

Occasionally on moist, rocky, north-facing slopes and canyons above drainages are stands of two evergreen broadleaf alliances: canyon live oak (*Q. chrysolepis*) forest and California bay (*Umbellularia californica*) forest. These trees, along with California black oak (*Q. kelloggii*), increase in extent as elevation rises. When joined by conifers such as ponderosa pine (*Pinus ponderosa*), incense cedar (*Calocedrus decurrens*), and even Douglas fir (*Pseudotsuga menziesii*), particularly in the northern foothills, stands have been called mixed evergreen forest (Sawyer et al. 1977).

The California black oak forest alliance is quite common along with ponderosa pine and incense cedar in the upper foothills and into the montane Sierra Nevada. California black oak stands, unlike the evergreen oaks, allow more light through their canopies. A variety of shrubs and herbs light up the understory, including California hop tree (*Ptelea crenulata*), California snowdrop bush (*Styrax redivivus*), deer brush (*Ceanothus integerrimus*), poison oak, toyon, and whiteleaf manzanita, as well as blue wild rye (*Elymus glaucus*), annual dogtail (*Cynosurus echinatus*), bromes (*Bromus* spp.), and melic grasses (*Melica* spp.).

Narrowly Distributed Woodlands

Patches of knobcone pine (*Pinus attenuata*) forest and McNab cypress (*Hesperocyparis macnabiana*) stands are scattered in the foothills, particularly on infertile serpentinite and gabbro soils. They occur as fire-adapted scrubby woodlands. When fire sweeps through stands, their closed cones are rapidly opened by the fire's heat, and their seeds rain down and germinate on bare soil.

Stands of McNab cypress are irregularly found in four counties in the foothills and adjacent Sierra Nevada ecoregion, mostly on private

lands, including River Pines in Amador County, Grass Valley in Nevada County, between Brownsville and Lake Mildred in Yuba County, and near Magalia and Concow in Butte County (Mallek 2009). Otherwise, this relatively rare alliance occurs in many scattered locations in the North Coast Ranges.

Other woodland alliances that are scattered from the southern to northern foothills include California juniper (*Juniperus californica*) woodlands, and in the upper elevations of the southern foothills, Jeffrey pine (*Pinus jeffreyi*) forest and singleleaf pinyon (*P. monophylla*) woodlands.

Grasslands and Vernal Pools

Grasslands abound across the region's rolling hillsides and valleys as glistening glades of green in the spring that turn golden yellow by summer and fall. To an undiscerning eye, they may all look the same with non-native grasses appearing overwhelmingly abundant. However, a dizzying array of grassland types occur depending on setting—from dry foothill slopes to wet swales and vernal pools. Myriad native herbaceous plants blanket the region and some even dominate in favorable rain years to the delight of their mutualistic pollinators. Fields with natives also regularly dominate on shallow soils and on harsher serpentinite and volcanic soils, while non-natives are especially dominant on deeper and richer soils.

Stands of native species also change in composition depending on moisture. For example, drier soils include the alliances fiddleneck (*Amsinckia [menziesii, tessellata]*) fields, California poppy (*Eschscholzia [californica]*) fields, popcorn

Top left: McNab cypress stands are scattered on serpentine in the foothills, here in Butte County with an understory of Brainerd's sedge (Carex brainerdii)*.*

Top right: Wildflowers delight the flats and slopes of serpentine soils in Mariposa County.

Above: Grassy swathes of soft brome (Bromus hordeaceus) *in deep-soil swales adjacent to vibrant goldfields and Sierra mock stonecrop* (Sedella pumila) *in shallower soils along Table Mountain in Fresno County.*
Julie M. Evens photos

flower (*Plagiobothrys nothofulvus*) fields, and annual brome (*Bromus* [*diandrus, hordeaceus*]–*Brachypodium distachyon*) grasslands, while the wetter soils include perennial rye grass (*Lolium perenne*) fields, common monkey flower (*Mimulus* [*guttatus*]) seeps, and various goldfield (*Lasthenia* spp.) and spike rush (*Eleocharis* spp.) alliances.

In the foothills, similar species replace one another geographically from north to south, reflecting shifting gradients in moisture, temperature, and soil. Replacement can be seen in the summer flowering tarweeds (*Holocarpha*), with

Above: Upland grassland assemblages include popcorn flower and fiddleneck with other wildflowers in the foothills of Fresno County.
Jennifer Buck-Diaz

Left: Wetland seeps of common monkey flower contrast the drier hillslopes of herbs along Highway 49 in Mariposa County.
Julie M. Evens

yellow flower tarweed (*H. virgata*) in the northern foothills, and Heerman's tarweed (*H. heermannii*) and San Joaquin tarweed (*H. obconica*) in the south. The needle grass species change from purple needle grass (*Nassella pulchra*) in the northern foothills to nodding needle grass (*N. cernua*) in the south. Geographic replacement in these and other common foothill genera such as *Clarkia*, *Madia*, and *Lupinus* are likely to be reflected in names for newly described plant associations as further analysis of the foothills vegetation patterns continues.

Other trends are visible on different

Above: Openings within oak woodlands exhibit glowing spring arrays of common madia (Madia elegans) *in the southern foothills.*
Jennifer Buck-Diaz

Right: Harvest brodiaea (Brodiaea elegans), *yellow flower tarweed, and other late-flowering annuals adorn the drying grasslands in summer.*
Julie M. Evens

Myriad wildflowers blanket the flat-topped table mountains and slopes along the foothills belt including goldfields, sky lupine, spikemoss (Selaginella hansenii), *and poppy at North Table Mountain Ecological Reserve near Oroville in Butte County.* Julie M. Evens

landforms. Volcanic tablelands are geologic features that contain species-rich wildflower fields in the foothills. These tablelands are basalt lava flows that filled in ancient river-beds of the Sierra Nevada many millions of years ago. While the surrounding surfaces around the rivers have eroded over time, the lava rocks are very resistant to weathering. Thus, the lava flows remain standing as tall, flat-topped table mountains in sinuous formations above the surrounding surfaces. These tabletops have colorful patches of vernal pool and swale vegetation intermixed with other shallow-soil native annual grassland types, including bright associations with goldfields (*Lasthenia gracilis, L. californica*), sky lupine (*Lupinus nanus*), and poppies (*Eschscholzia* spp.). Immediately below the tabletops, California buckeye stands form in narrow bands clinging to the steep rocky sides along the tables' edges or in draws coming down their slopes. You can find table mountains near Oroville in Butte County, near Jamestown and Sonora in Tuolumne County, and near Auberry (above and below the San Joaquin River) at Kennedy Table Mountain in Madera, and the adjacent Big and Little Table mountains in Fresno County. Other vernal pools, expansive swales, and sheets of wildflowers occur on volcanic soils in the Lassen foothills, with similar plant assemblages at Hog Lake and Dales Lake in Tehama County.

Above: Large vernal pool features and related grasslands occur in the Lassen Foothills on volcanic soils such as at Hog Lake in Tehama County.

*Left: Endemic vernal pools occur along the lava-flow table mountains, including goldfields, annual checkerbloom (*Sidalcea calycosa*), rosy Douglas' meadowfoam (*Limnanthes douglasii *ssp.* rosea*), and white-tip clover (*Trifolium variegatum*) near Prather in Fresno County.*

Julie M. Evens photos

Above: Reservoirs, including Lake McClure along the Merced River in Mariposa County, provide significant water to the foothills and valley while riverside habitat has dramatically declined. Julie M. Evens

Riparian and Wetland Vegetation

Waterways have eroded down valleys and deeply incised canyons of the Sierra Nevada, forming a mighty network of rivers and streams with majestic forests and dense riparian vegetation. Most are westerly flowing and change from rapid to more slow-moving as they reach the foothills. Reservoirs for agricultural irrigation, flood control, and water supply have become common features, especially in the foothills, before reaching the fertile soils of the Great Valley. While dams and reservoirs have modified natural flooding in all but one river, the Cosumnes, the vegetation is still quite intact above them.

Riparian woodland stands of white alder (*Alnus rhombifolia*), Oregon ash (*Fraxinus latifolia*), California sycamore (*Platanus racemosa*), and Fremont cottonwood (*Populus fremontii*), along with shrubby stands of mulefat

Left: Riparian vegetation and streams provide habitat for native fish, such as this wild rainbow trout (Oncorhynchus mykiss) *in the Yuba River.* Roger Bloom

(*Baccharis salicifolia*), sandbar willow (*Salix exigua*), black willow (*S. gooddingii*), and red willow (*S. laevigata*), occur regularly in drainages with permanent and seasonal water flows. Certain associations of interior live oak, California buckeye, and valley oak alliances also occur in intermittently to seasonally flooded stream channels and terraces in the region.

Less common riparian vegetation includes stands of button willow (*Cephalanthus occidentalis*) where water seasonally pools at stream and lake edges, California coffee berry (*Frangula californica*) scrub in intermittent streamsides and rocky outcrops, and torrent sedge (*Carex nudata*) patches in regularly scoured rocky streams. Unfortunate invasions of tree of heaven (*Ailanthus altissima*) and edible fig (*Ficus carica*)

Above: Intact riparian vegetation is quite complex in the foothills, including California sycamore, white alder, western redbud, and various willows, as found in Tehama County.
Julie M. Evens

Left: Torrent sedge patches occur regularly along scoured streams and serpentine creeks in the foothills region, including this small creek in Tuolumne County.
Sau San

Top: Whiteleaf manzanita, chamise, and wedge leaf ceanothus are the most common chaparral types in the foothills.
Julie M. Evens

*Above: Chaparral with unusual assortments of plants occur on the gabbroic soils of the Pine Hill area of western El Dorado County, including rare endemic plants found nowhere else in the world such as Pine Hill ceanothus (*Ceanothus roderickii*).*
Jeff Bisbee

also occur in riparian zones of the foothills. Land managers should take care to remove these invaders before they become major problems.

Other wetland vegetation is similar to that of the Great Valley, including alliances of creeping rye grass (*Leymus triticoides*), sedges (*Carex* spp.), rushes (*Juncus* spp.), cattails (*Typha* spp.), and bulrushes (*Schoenoplectus* spp.). They can be found in low pockets along stream oxbows and terraces, at pond edges, in marshes, and other low-lying features. More seasonally or intermittently flooded vegetation includes Spanish clover (*Lotus unifoliolatus* var. *unifoliatus*) fields, deer grass (*Muhlenbergia rigens*) beds, and smartweed–cocklebur (*Persicaria lapathifolia* –*Xanthium strumarium*) patches.

Chaparral and Other Shrublands

Chaparral forms open to dense bands of vegetation on slopes and ridges in the foothills. The most regularly distributed chaparral alliances are wedge leaf ceanothus and whiteleaf manzanita, common in drier settings and on basalts and serpentinite substrates across the northern foothills. Stands dominated by chamise (*Adenostoma fasciculatum*) and mixes of chamise and wedge leaf ceanothus are more common in drier settings of the central and southern foothills.

Unusual associations of whiteleaf manzanita, chamise, and creeping sage (*Salvia sonomensis*) occur with a suite of rare plants on gabbro soil of the Pine Hill region, between Folsom Lake and Highway 50 in western El Dorado County. The reddish soil is rich in iron and magnesium, and like serpentinite, is nutrient-poor and difficult for typical plant growth. A remarkable collection of eight rare plants have adapted to grow upon this unusual rock type. Unfortunately, this region is interlaced with suburban and commercial development, and current conservation efforts have been successful in protecting only 10% of this 14,000-hectare area.

On the cooler, north-facing slopes and draws, chaparral scrub of interior live oak, scrub oak (*Quercus berberidifolia*), and birch leaf mountain mahogany (*Cercocarpus montanus*) alliances occur. Pockets of California buckeye and western redbud (*Cercis occidentalis*) often intermix in these cooler settings. Scattered stands of Brewer oak (*Q. garryana* var. *fruticosa*) scrub

alliance occur on sheltered slopes and below steep vertical slopes in higher elevations of the foothills.

In lower to mid elevations across the foothills, California yerba santa (*Eriodictyon californicum*) and flannelbush (*Fremontodendron californicum*) stands occur in areas with recent fire or other disturbances, especially on exposed sandy slopes and flats. Other recently disturbed slopes can be laden with stands of poison oak scrub alliance, especially on rocky, moist north or neutral slopes where oak forests and woodlands previously occurred. In upper-elevation forested areas, chaparral whitethorn (*Ceanothus leucodermis*) and deer brush chaparral appear in areas with recent fire, logging, or other disturbance.

The central and southern foothills have a diversity of other shrublands, including California yerba santa and

Top: Brewer oak scrub in fall color below steep slopes in the southern foothills.
John Menke

Above: Post-burn and disturbance shrublands of California yerba santa and poison oak occur regularly in the foothills.
Julie M. Evens

Above: Southern foothills vegetation takes on characteristics of both coastal and desert alliances, including California buckwheat with an assortment of wildflowers along the lower slopes near Gorman in northern Los Angeles County.
JULIE M. EVENS

Below: Rubber rabbitbrush shrubland along a lower exposed slope in southern Kern County.
CNPS PHOTO

bush poppy (*Dendromecon rigida*) scrub in random pockets of recently burned chaparral, which often transitions to chamise. Silver bush lupine (*Lupinus albifrons*) scrub, deer weed (*Lotus scoparius*) scrub, and narrowleaf goldenbush (*Ericameria linearifolia*) scrub alliances occur on steep rocky slopes or along terraces and floodplains of streams.

In the southernmost foothills, chaparral and related vegetation include mixtures of desert-like and coastal vegetation types. Cup leaf ceanothus (*Ceanothus greggii*), big sagebrush (*Artemisia tridentata*), allscale (*Atriplex polycarpa*), Wright's buckwheat (*Eriogonum wrightii*), and rubber rabbitbrush (*Ericameria nauseosa*) scrub alliances are more representative of the desert edge. California buckwheat (*Eriogonum fasciculatum*) scrub, thick leaf yerba santa (*Eriodictyon crassifolium*) scrub, oak gooseberry (*Ribes quercetorum*) thickets, and Tucker oak (*Quercus john-tuckeri*) chaparral alliances are more representative of the south and central coast. Scale broom (*Lepidospartum squamatum*) scrub alliance also occurs in sandy or gravelly washes and stream terraces of the southern foothills.

One of the most rare and endemic shrubland types is bush anemone (*Carpenteria californica*). It occurs as chaparral on moist slopes with California buckeye, interior live oak, and western redbud, occurring only between the Kings and San Joaquin rivers (west and south of Shaver Lake). Sierra Foothills Conservancy's Miller Preserve at Black Mountain has afforded protection of some of these stands.

Conservation and Restoration

The foothills support a diversity of wildlife, including many migratory and resident birds and mammals in the upland habitats and lower watersheds, as well as a host of resident and anadromous fish. However, river diversions, rapid expansion of urban growth, and other disturbances have seriously impacted and fragmented the landscape. Historic and current mining activity has also significantly altered areas of the foothills, including mining for clay where endemic species occur (e.g., Ione manzanita), for gravels and gold in riparian areas, and for gold and other minerals on hillsides where chaparral, oaks, and grasslands have begun to reclaim their former extent. Where dams and diversions have altered flow regimes, blocked fish passage, and damaged riparian areas, existing facility managers and agencies should explore options for allowing streams to flow as naturally as possible.

Since people and vineyards have moved into the region in scattered and concentrated developments, they have limited the ability of fires and wildlife to naturally move across the landscape, are drawing down wetland and riparian resources, have cleared natural vegetation, and are causing erosive effects. Future development should be carefully planned to limit impact to as little area as possible. Hopefully, larger ranchlands will continue to promote habitat diversity with carefully managed livestock grazing without degradation of riparian, grassland, and other critical areas. Continued urban growth, changes in climate and in historic fire regimes, and spread of invasive species threaten the existence of some vegetation types and assure changes in composition for others. Current die-off of blue oak, interior live oak, foothill pine, and ponderosa pine trees could be part of a natural cycle of drought coupled with parasite and insect damage, yet may also be an effect of changing climates. However, we still have time to act wisely to preserve and restore the remaining wildness for future generations to cherish.

Julie M. Evens

Above: Stands of bush anemone, rare in the central foothills, bloom here at Miller Preserve at Black Mountain in Fresno County.
Aaron Wentzel

Inset: Bush anemone.
Tom Zimoski

Below: Oak titmouse (Baeolophus inornatus), *a bird of conservation concern, prefers open oak woodlands and lives year round west of the Sierra Nevada in California.*
John Sterling

CHAPTER EIGHT

Great Valley

Anyone who has flown up or down California and looked out over the Great Valley could be forgiven for thinking that what was once a vast expanse of wildflowers is now a sea of agriculture covering the entire valley floor from the Sierra Nevada foothills to the western edges of the Sacramento and San Joaquin Valleys. It is true that most of the historic vegetation of the Great Valley has been urbanized, converted to agricultural land, and substantially altered by grazing and an influx of exotic grasses and forbs. Nevertheless, large areas of uncultivated lands remain: at the northern end of the Sacramento Valley; along the terraces bordering the eastern edge of the Great Valley; at the edges of the Delta such as Jepson Prairie; in the San Joaquin and Tulare basins; and, in the Antelope Plain and Elk Hills of the southwestern San Joaquin Valley. In addition, many smaller remnants of natural vegetation are scattered throughout the region.

Top: Almond trees in bloom in the Great Valley.
GARRY HAYES

Above: Grasslands in the Great Valley can display colorful wildflowers.

Opposite: Vast alluvial terraces remain intact with grasslands and riparian woodlands at Tejon Ranch along the southern San Joaquin Valley.
MICHAEL WHITE PHOTOS

Plant communities and major habitats of the Great Valley fall into five general categories: grasslands, vernal pools, alkaline habitats, riparian habitats, and freshwater marsh. Most of these are characterized by herbaceous species—woody vegetation is primarily associated with rivers and streams. In addition, other tree- and shrub-dominated communities spill down onto the valley floor in a transition zone with foothill oak woodlands and desert scrub communities. The Great Valley plant communities are

Above: Vernal pool in eastern Sacramento County with Fremont's goldfields and white-tip clover.
GLEN HOLSTEIN

*Left: Transitional scrub of bladderpod (*Cleome isomeris*) with annual grasslands overlooking the southern San Joaquin Valley.*
DAVID MAGNEY

generally influenced by climate and topography, but the primary factors determining the composition of plant associations are soil properties, primarily pH, texture, and the availability of water.

The Great Valley's climate is usually Mediterranean, although the southern end of the valley is more accurately described as semi-arid desert (Germano et al. 2011). Summers are hot, and measurable precipitation during June, July, and August is rare. Summer temperatures are often the most extreme at the north and south ends of the Great Valley, with offshore delta breezes moderating the mid-valley highs. Winters are cool, but hard freezes are rare, and the growing season is essentially year-round. Most of the precipitation, which falls primarily as rain, comes between mid-October and mid-March. Rainfall varies across the Great Valley, being lower in the rain shadow of the Coast Ranges than along the east side of the valley, and decreasing from north to south. The northern Sacramento Valley receives 500–600 mm of precipitation annually. In contrast, annual rainfall along the western edge and southern end of the San Joaquin Valley drops to 125–150 mm, which is more typical of deserts.

In his book *Assembling California*, John McPhee (1993) describes the Great Valley as "far more planer than the plainest of the plains" and "very much flatter than Iowa or Kansas." This flatness is accentuated where land leveling has further smoothed out the micro-topography. Toward the valley margins, however, flatness gives way to gently rolling hills where mountain-born streams have deposited alluvial fans and terraces that have been tilted upward by tectonic processes. These terraces, which vary from a few kilometers to tens of kilometers wide and are broader along the east side of the valley, possess examples of our oldest soil types, some of which have a well-developed claypan or hardpan. More prominent topographic relief occurs in the south end of the valley at the Elk Hills and the grass-covered foothills northeast of Bakersfield. The Sutter Buttes, a small, isolated volcanic range, are the most prominent feature of the Sacramento Valley.

The association between soil type and vernal pool plant communities has long been recognized (Holland 1978;

Top: An aerial view of the Sutter Buttes, a prominent topographic feature near the northern end of the Great Valley.
Todd Keeler-Wolf

Above: Saltbush scrub, an alkaline tolerant alliance, in the San Joaquin Valley.
Robert Preston

Above: Wetlands and riparian waterways of the Sacramento-San Joaquin River Delta.
GARRY HAYES

Below: Soft bird's beak (Cordylanthus mollis *ssp.* mollis)*, an endangered plant, grows in marshes of pickleweed* (Sarcocornia pacifica) *and salt grass; Hill Slough, near Jepson Prairie.*
ROBERT PRESTON

Barbour et al. 2003), but the influence of soils on other Great Valley plant communities is less well-documented. Alkaline and other salt-rich soils that are abundant in or adjacent to the basins that formerly occupied the middle reaches of the Great Valley profoundly affect plant growth and support a suite of alkaline- and salt-tolerant shrubs and herbs (Preston 2010). Some soil types are quite uncommon, such as the eolian sand deposits at the Arena Plains that support several rare or disjunct species (Silveira 2000). Although a correlation exists between some plants and clay soils, plant associations occurring on clay soils have yet to be investigated.

Wetlands and riparian communities historically were an order of magnitude more extensive than at present but still can be found throughout the Great Valley. These communities add species richness to the Great Valley and they provide highly valuable wildlife habitat.

GRASSLANDS

The traditional paradigm of a pristine California prairie dominated by purple needle grass (*Nassella pulchra*) continues to shape perceptions about the original vegetation of California and guide the direction of research on the ecology and restoration of native perennial grasslands in California (Bartolome et al. 2007). However, an alternative point of view on the vegetation of the Great Valley has been gaining wide acceptance. Jepson (1925) proposed that the original vegetation of California's Great Valley was not prairie but consisted of vast areas dominated by native herbaceous annuals interspersed with alkaline flats (characterized by alkali-tolerant species) and vernal pools. Historical accounts have supported this point of view, providing a narrative of vegetation composed primarily of annuals, rather than bunchgrasses, with scattered shrubs occurring in xeric areas (Wester 1981; Schiffman 2007; Minnich 2008). A recent analysis of phytoliths (microscopic particles of silica found in plant cells) in soil samples from widespread Great Valley grassland sites found that many of these grasslands are of recent origin, which is the first direct evidence supporting Jepson's hypothesis (Evett and Bartolome 2013). Purple needle grass grasslands, where present, occurred in more mesic habitats such as coastal prairies and in the foothills adjacent to the Great Valley, but they were found only at scattered locations in the Great Valley proper. In addition, grassland communities dominated by species other than purple needle grass may also have been present, such as creeping rye grass (*Leymus triticoides*) turfs (Holstein 2001).

Grasslands occur in all parts of the Great Valley, although the most extensive and unfragmented areas are located around the valley margins. In the central portions of the valley, grasslands are present as small fragmented stands or, where more abundant, are interspersed with alkaline habitats, particularly in the San Joaquin Valley. The grasslands at Jepson Prairie are well known for their vernal pools as well as extensive stands of purple needle grass grassland.

Grasslands of the Great Valley are often perceived as dominated by non-native grasses, which has led to their classification as non-native grassland (Holland 1986) or California annual grassland (Sawyer and Keeler-Wolf 1995). However, the relative cover of non-native grasses and native forbs changes during the growing season and also varies from year to year as a result of climatic variation, and in some years and locations spectacular wildflower blooms belie the grassland designation. This focus on the non-native elements of the grassland flora obscures the fact that they harbor

Top: Creeping rye grass turf may have been a prominent early California grassland vegetation type.

Above: Annual brome grassland creates a vegetation mosaic with blue oak woodlands.

Todd Keeler-Wolf photos

Above: Fiddleneck field at Tejon Ranch in the southern San Joaquin Valley.
Betsy Harbert

Below: Northern harrier (Circus cyaneus), *a California bird species of special concern, forages on the wing and nests on open ground.*
John Sterling

a tremendous diversity of native forbs and grasses—and at the same time may create a bias that diminishes their perceived conservation value. A more useful approach may be to recognize vegetation types based on their characteristic or diagnostic native or non-native species. For a discussion of the changing perspective on California grasslands, see Holstein (2011) and "Changes Since the First Edition" in Sawyer et al. (2009).

The most common and familiar semi-natural herbaceous stands are annual brome (*Bromus* [*diandrus, hordeaceus*]–*Brachypodium distachyon*) grasslands and wild oats (*Avena* [*barbata, fatua*]) grasslands. In the arid southern San Joaquin Valley, red brome or Mediterranean grass (*Bromus rubens–Schismus* [*arabicus, barbatus*]) grasslands are common. However, the historic native species alliances and associations are still present and have persisted, despite the dominance by non-native grasses (Keeler-Wolf et al. 2007). Focused vegetation sampling in the San Joaquin Valley and Carrizo Plain has recently identified many new grassland alliances and associations (Buck-Diaz et al. 2013). Some of the alliances are characterized by conspicuous wildflowers, such as fiddleneck (*Amsinckia* [*menziesii, tessellata*]) fields, California poppy (*Eschscholzia*

[*californica*]) fields, and popcorn flower (*Plagiobothrys nothofulvus*) fields. Undoubtedly, many more alliances will be delineated as closer attention is paid to some of the more localized plant communities, such as wildflower fields that occur on shallow soils or in areas with high concentrations of geophytes, herbaceous perennials that die back during the summer drought and survive underground as rhizomes, bulbs, or corms (*Brodiaea*, *Dichelostemma*, *Triteleia* and other mostly monocot genera).

Above: Sky lupine (Lupinus nanus) *and poppy often intermix with popcorn flower fields at Tejon Ranch.*
Michael White

Right: Giant kangaroo rat (Dipodomys ingens), *an endangered species, is primarily a grass and forb seed-eater.*
Christine Van Horn Job

Above: Fremont's tidy-tips vernal pool alliance in a shallow pool at the Arena Plains Unit of Merced National Wildlife Refuge.
Gary Zahm

Right: Popcorn flower field at Tejon Ranch.
Todd Keeler-Wolf

Opposite top: Fremont's goldfields is the most common and showy vernal pool vegetation type.
Gary Zahm

Opposite right: Shallow soils provide high concentrations of geophytes such as Brodiaeas that will likely be defined in associations as more work is done in this region.
Julie M. Evens

Vernal Pools

Much has been written about California grasslands, yet even more has been written about vernal pools, which are one of the jewels in California's crown. Vernal pools are a natural and obvious habitat type, yet classifying the vegetation has proven to be a difficult task. Vernal pool classification has traditionally been based on geographic location, landscape position, the presence of regionally rare species, and soil types (Holland 1978). Seventeen vernal pool regions have been identified throughout California, five of which are represented in the Great Valley region (Keeler-Wolf et al. 1998). More recently, ecologists have begun to classify and name vernal pools by the plant associations they contain (Barbour et al. 2007). Vernal pools are typically

Top left: White-tip clover swale at the Flying M Ranch in Merced County.
Julie M. Evens

Top right: Pale spike rush marsh alliance growing in a deeper pool on Tuolumne Table Mountain.
Todd Keeler-Wolf

Above: Spinescale (Atriplex spinifera) scrub growing along the alkaline soils of Semitropic Ecological Preserve in Kern County.
Richard Spjut

complexes of two or more plant associations, most of which are limited in distribution to a single vernal pool region.

Vernal pool alliances are named according to the characteristic species present. The most common and showy vernal pool type is the Fremont's goldfields–downingia (*Lasthenia fremontii–Downingia* [*bicornuta*]) vernal pools alliance, which is typical of the shallow portions of vernal pools. The Fremont's tidy-tips–blow wives (*Layia fremontii–Achyrachaena mollis*) vernal pools alliance is often found in very shallow pools or on pool margins. The white-tip clover (*Trifolium variegatum*) swales alliance is found in swales but also on pool margins. The deeper portions of vernal pools, which are often less species-rich and not as showy, are represented by the smooth goldfields (*Lasthenia glaberrima*) vernal pool bottoms and pale spike rush (*Eleocharis macrostachya*) marsh alliances. Some vernal pool alliances are rare. For example, the California button-celery (*Eryngium aristulatum*) patches alliance is only found in claypan pools along the western edge of the Sacramento Valley, and the water blinks–annual checkerbloom (*Montia fontana–Sidalcea calycosa*) vernal pools alliance is restricted to volcanic table mountains.

Alkaline Habitats

Alkaline habitats are generally located in or adjacent to the basins that occupy the middle reaches of the Great Valley. They range from Glenn and Butte Counties in Sacramento Valley, to the southern end of San Joaquin Valley in Kern County. Similar habitats are present in the valleys of the adjacent Central Coast Ranges, such as the Carrizo Plain, the Cholame Valley, and the Livermore Valley. Much of the land formerly occupied by these basins has been converted to agriculture. Soil surveys published by the National Resources Conservation Service, especially the older ones, provide detailed prescriptions for "reclaiming" alkaline soils by adding soil amendments such as gypsum (calcium sulfate) or sulfur. Much of the remaining habitat where alkaline soils can be found is located on state and federal lands. The largest areas of remaining alkaline habitat are within the National Wildlife Refuges (NWR), including the Sacramento, San Luis, and Kern NWR complexes. Small areas of habitat are also preserved in California Department of Fish and Wildlife Ecological Reserves, including Alkali Sink, Kerman, and Allensworth Reserves.

Very little historic information is available on the species composition or ecology of Great Valley alkaline habitats. The first systematic surveys of alkaline habitats were done by Joseph Burtt Davy (1898), who recognized several plant communities on alkaline soils around lake basins in the San Joaquin Valley. He found that these communities could be generally zoned according to moisture availability. Saltbush (*Atriplex* spp.) scrubs occurred on the higher, more xeric sites on the valley margins, and iodine bush (*Allenrolfea occidentalis*) scrub occurred in the basins, usually in the presence of a high water table. At intermediate sites he recognized three plant communities: alkali goldenbush (*Isocoma acradenia*) scrub; bush seepweed (*Suaeda moquinii*) scrub; and alkali sacaton (*Sporobolus airoides*) grassland. Alkali sacaton grassland at the San Luis NWR and at Great Valley Grasslands State Park encompasses one of the largest areas of native perennial grassland in the Great Valley

Top: Iodine bush scrub can be found in areas with high water tables.
Todd Keeler-Wolf

Above: Extensive tar plant fields can be found in alkaline habitats during late summer.
Robert Preston

(Barry 1972; Solomeshch and Barbour 2006). Salt grass (*Distichlis spicata*) flats are often interspersed with these other communities.

The herbaceous vegetation of alkaline habitats is often dominated by the ubiquitous non-native grasses, but halophytes are common. In early spring, common goldfields (*Lasthenia gracilis*) provides extensive wildflower displays where alkaline or saline soils suppress the growth of grasses. In later summer, tar plant (*Centromadia [pungens]*) fields form extensive stands. Within the annual grassland matrix, alkaline rain pools and salt scalds form a network within the uplands, with alkaline vernal pools occurring in swales and depressions. The vernal pool community characteristic of alkaline soils is the Fremont's goldfields–salt grass alkaline vernal pools alliance. These pools contain a mix of widespread vernal pool endemics as well as halophytes, including alkali weed (*Cressa truxillensis*), alkali heath (*Frankenia salina*), and annual saltbush species. The pool margin community is often composed of low barley (*Hordeum depressum*) grassland. Playa pools, alkaline rain pools, and salt scalds have highly alkaline or saline soils and generally lack vegetation except on the margins, where barley grassland, tar plant fields, or a community dominated by halophytes may occur. The bare areas of playa pools and alkaline rain pools often develop a cryptogamic crust—a unique community of mosses, liverworts, lichens, fungi, algae, and bacteria that live on the soil surface (Preston 2010).

Top: Fremont's goldfields–saltgrass alliance contains both vernal pool species and halophytes.
Ayzik Solomeshch

Left: A salt scald in alkaline grassland where vegetation is unable to grow.
Robert Preston

Riparian Habitats

Formerly, riparian habitats were extensive along the rivers and streams of the Great Valley. Historic accounts describe the extent of riparian habitat along the rivers and comment on the great size of the valley oak (*Quercus lobata*) trees. W. H. Brewer described his party's long, weary trek in the early 1860s across the treeless San Joaquin Valley, finally finding relief at the oak woodlands along the floodplains of the Kings and Kaweah Rivers (Brewer 1966). Remnants of these historic woodlands are still present at Mooney Grove Park and the Kaweah Oaks Preserve. However, channelization and the encroachment of agriculture have reduced the extent of riparian habitats to only 11–13% of the historic extent (Katibah 1984), and they now mostly occur as thin strips along streams and man-made channels. Despite this loss, riparian habitats are still fairly widespread in the Great Valley, and fine examples of Great Valley riparian woodlands can be observed at Sacramento River Wildlife Area, Bidwell Park, Cosumnes River Preserve, and Caswell Memorial State Park. For a comprehensive review of the management history and vegetation ecology of Great Valley riparian habitats, see Vaghti and Greco (2007).

The characteristic riparian alliance along the Great Valley's rivers and streams is Fremont cottonwood (*Populus fremontii*) forest, a dense forest community with well-developed tree and shrub canopies. Because Fremont cottonwood

Above: Fremont cottonwood forest along the banks of Thomes Creek.
Todd Keeler-Wolf

*Below: A rookery of double-crested cormorants (*Phalacrocorax auritus*) atop a cottonwood tree.*
John Sterling

Top left: Valley oak woodland on the flats adjacent to the more active flood plain.
GARY ZAHM

Top right: Swainson's hawk (Buteo swainsoni)*, a threatened species, forages and nests peripheral to riparian areas.*

Above: Yellow warbler (Setophaga petechia) *is a riparian-dependent bird species of special concern in California.*

Right: White-breasted nuthatch (Sitta carolinensis) *is a permanent resident in oak woodlands and oak-containing riparian corridors.*
BIRD PHOTOS: JOHN STERLING

requires disturbance for establishment, this community occurs primarily within the active floodplain. In addition to Fremont cottonwood, other trees that may dominate the canopy include oaks (*Quercus* spp.), box-elder (*Acer negundo*), Oregon ash (*Fraxinus latifolia*), northern California black walnut (*Juglans hindsii*), California sycamore (*Platanus racemosa*), black willow (*Salix gooddingii*), and red willow (*S. laevigata*). The shrub canopy is dominated by various shrubs, including button willow (*Cephalanthus occidentalis*), poison oak (*Toxicodendron diversilobum*), and arroyo willow (*S. lasiolepis*). Lianas of California wild grape (*Vitis californica*) are often conspicuous, climbing upward into and over riparian trees toward accessible light.

Valley oak woodland also occurs along rivers and streams of the Great Valley but extends further away from the active floodplain. The composition of the valley oak woodland plant community is similar to that of Fremont cottonwood forest, but the tree canopy is often more open, limiting shade-tolerant species. An herbaceous layer is often present, dominated by annual brome grasses or by native creeping rye grass.

White-root (*Carex barbarae*) beds are also a component of the herbaceous layer at some locations.

California sycamore woodlands occur along streams in the Coast Range and Sierra Nevada foothills, sometimes extending into the adjacent valley. Because these woodlands occur along intermittent or seasonal streams, the habitat is typically more arid than other

Above: California sycamore woodlands occasionally occur in the Great Valley but are more common in the Sierra Nevada foothills, such as here along Bidwell Creek.

Right: Sandbar willow thicket adjacent to Fremont cottonwood forest along the Sacramento River.
TODD KEELER-WOLF PHOTOS

riparian woodlands. The tree canopy is relatively open and the understory canopy is often sparse or composed of scattered mulefat (*Baccharis salicifolia*) or willows. The herbaceous layer is generally dominated by introduced annual grasses. California sycamore woodland is one of the Great Valley's rarest alliances (Sawyer et al. 2009).

Riparian shrubland alliances are also widespread along the Great Valley.

Above: Button willow thicket along the Sacramento Delta.
TODD KEELER-WOLF

Left: Himalayan blackberry bramble along disturbed edges of the Sacramento River.
JOHN O. SAWYER

A dense California rose briar patch with scattered patches of hardstem bulrush along a bank of the Sacramento River Delta. TODD KEELER-WOLF

Shrubland alliances typically occur on stream banks or in the active floodplain, and the same species often form the shrub layer of riparian woodlands. The distribution of each of these alliances is strongly influenced by the disturbance regime. Sandbar willow (*Salix exigua*) thickets and mulefat scrub are present within channels and along banks at locations that are subject to active scour, as both are colonizers of recently disturbed areas. Button willow thickets occur in narrow strips along stream banks and levees that are seasonally flooded but not scoured. California rose (*Rosa californica*) briar patches, which occur on stream banks and terraces, can extend into the drier portions of the riparian zone, due to a greater level of drought tolerance. Blue elderberry (*Sambucus nigra*) stands also occur on stream banks and terraces and can extend into uplands outside of the riparian zone. Blue elderberry is shade-intolerant, and persistence of stands in the riparian zone is dependent on episodic, albeit infrequent, disturbance. Himalayan blackberry (*Rubus armeniacus*) brambles are common in disturbed areas in the riparian zone but are also invasive in riparian habitat, having similar ecological requirements as California rose.

Freshwater Marsh

Historically, the large, low-lying areas of the central trough of the Great Valley

Top: Hardstem bulrush marsh, northern San Joaquin Valley.
Gary Zahm

Above: Lesser snow geese (Chen caerulescens caerulescens)*, flourishing in wintering areas such as Sutter National Wildlife Refuge in the Great Valley, along the Pacific Flyway.*
John Sterling

hosted vast reaches of freshwater marsh. The principal marsh areas were centered on the overflow basins of the Sacramento Valley (Butte, Colusa, Sutter, American, and Yolo) and the San Joaquin Valley (San Joaquin, Tulare). Following passage of the Swamp and Overflowed Lands Act of 1850, a concerted effort was made to "reclaim" wetlands for agriculture and to protect farmland and the public from flooding. A vast network of levees and canals was built to drain the wetlands and divert river flows to support agriculture, and nearly all of the former marsh lands of the Sacramento and San Joaquin Valley basins were converted to crop lands. The movement to preserve and restore wetlands in the Great Valley began with the recognition of their importance to fish and waterfowl, and then gained considerable traction with

the rise of the environmental movement and the establishment of Federal wetlands policies and regulations. Today, extensive areas of managed freshwater marsh are found primarily within the National Wildlife Refuge system and state wildlife management areas. Other small areas of freshwater marsh are scattered throughout the Great Valley, often associated with rivers, streams, ponds, lakes, and other water bodies. Philip Garone (2011) has written a fascinating history of the wetlands of the Great Valley, in which he describes the roles of politics, public policy, and the evolution of public perceptions in determining the current state of these wetlands.

The characteristic freshwater marsh community of the Great Valley is the tule marsh, dominated by one or more species of tule or bulrush (*Schoenoplectus* spp.).

Top: Semi-natural perennial pepper weed patch near the Sacramento Delta.
Todd Keeler-Wolf

Above: Cattail commonly occurs in marshes.
Gary Zahm

Top: Common reed marsh, an invasive wetland type found in the Sacramento River Delta.
Todd Keeler-Wolf

*Above: A hoverfly (*Eupeodes *sp.) bee-mimic visiting a vernal pool endemic, flatface calicoflower (*Downingia pulchella*), at Glide Tule Ranch of the Yolo Bypass Wildlife Area.*
David Rosen

*Opposite: Protection of wetlands in wildlife refuges provides habitat to many wildlife species such as the migrating sandhill cranes (*Grus canadensis*) at Cosumnes River Preserve, protected and managed by The Nature Conservancy, Bureau of Land Management, Ducks Unlimited, California Department of Fish & Wildlife, among others.*
Robert Wick

The most common types of tule marsh are hardstem bulrush (*S. acutus*) marsh and California bulrush (*S. californicus*) marsh. Open water between the tules may be dominated by emergent or floating aquatics, such as mosquito fern (*Azolla filiculoides*), duckweed (*Lemna* spp.), water primrose (*Ludwigia peploides*), the rare Sanford's arrowhead (*Sagittaria sanfordii*), or the exotic water hyacinth (*Eichhornia crassipes*). Cattail (*Typha* [*angustifolia*, *domingensis*, *latifolia*]) marshes are another common freshwater marsh community in the Great Valley. Composition of marsh vegetation is largely dependent on the depth and duration of inundation. Along the marsh margins or in other shallow or temporarily inundated areas are Baltic and Mexican rush (*Juncus* [*arcticus* ssp. *littoralis*, *mexicanus*]) marshes, and smartweed–cocklebur (*Persicaria lapathifolia*–*Xanthium strumarium*) patches. Because marshes are frequently managed or disturbed, communities dominated by invasive non-native species can be present, such as perennial pepper weed (*Lepidium latifolium*) patches. Although common reed (*Phragmites australis*) marshes are native to parts of California, it may be invasive in disturbed marshes of the Great Valley.

Vegetation Persistence and Conservation

In California, water is sacred. There has been a much greater emphasis on protecting vernal pools, riparian habitat, and freshwater marsh than on conserving terrestrial habitats. Much of the protection to plant communities has been incidental to preserving the ecological services provided to wildlife species. Protection of freshwater marshes in national wildlife refuges and state wildlife management areas has been motivated largely by the desire to manage wetlands as feeding habitat for waterfowl—to encourage them to avoid agricultural lands, or to provide hunting opportunities.

Vernal pool conservation has been driven by Clean Water Act provisions that require mitigation for the loss of wetlands, either through preservation, creation, or both. Riparian communities are important components of fish habitat because they provide both shade and input of organic materials, and much of the preservation and restoration of riparian woodland and forest is directed at enhancing habitat for anadromous fish species.

In contrast, there has been little focus on the conservation of grassland habitat, except where it is regarded as watershed for vernal pools. Because of the tremendous loss of Great Valley habitats and the continuing loss of upland habitats, there is need for a conservation ethic that recognizes a broader perspective on the value of these lands as habitat both for animals and plants, rather than a narrow focus on managing a set of specific ecological functions.

Robert E. Preston

CHAPTER NINE

Central California Coast Ranges

Top: The gentle slopes of Panoche Hills, in the drier part of the Diablo Range.
Robert Wick

Above: Tesla Park with blue oak and coastal scrub, in the moister part of the Diablo Range.

Opposite: Woodlands, chaparral, and grasslands along the inner Coast Ranges.
Pete Veilleux photos

The Central Coast Ranges are physiographically intermediate between the steep, rugged mountains of the immediate coast to the west, the outer Coast Ranges, and the exceptionally flat topography of the San Joaquin Valley to the east. The region presents a series of long, northwest–southeast trending valleys interspersed with elongated ridges parallel to, but inland from, the Central California coastline. Rainfall totals tend to decrease inland from the western edge, while summer temperatures tend to increase. Moderate coastal maritime air masses regularly slide over the low passes and the occasional gaps in these ridges to provide cool and dependable breezes, powering wind farms at Altamont Pass and near the Carquinez Straits.

The inner Coast Ranges are neither as steep nor as rugged as the outer Coast Ranges. At the northernmost point of the region are the heavily urbanized Contra Costa (Berkeley) Hills where cooling summer fog regularly pours over them, moderating the climate for some of the most inland stands of coast redwoods in Redwood Regional Park. The first

Above: California sycamore alluvial woodland along Arroyo Seco, Monterey County.
Steve Perdue

Opposite top: Semi-desert species in the Temblor Range above the Carrizo Plain, San Luis Obispo County.
Robert Wick

Opposite bottom: Pinnacles National Park in eastern Monterey and western San Benito counties preserves quintessential Central Coast Range habitats, including chaparral, coastal scrub, meadows, oak and pine woodlands, and rock outcrops.
Julie M. Evens

major mountain range to the southeast is the Diablo Range. It extends more or less continuously from San Pablo Bay and the Carquinez Straits to Polonio Pass and then into northeastern San Luis Obispo County. The Diablo Range is often subdivided (from north to south) into the Contra Costa Hills, Mount Diablo, the Hamilton Range, the Panoche Hills, the San Carlos Range, and the Estrella Hills. To the west, the Gabilan Range runs parallel to the Diablo Range but does not connect to it. This range separates the Salinas Valley from the more interior San Benito Valley, the southern extension of the much larger and heavily urbanized Santa Clara Valley. The Santa Clara Valley is framed by the Santa Cruz Mountains to the west and by the Diablo Range to the east.

The Diablo Range contains large areas of wilderness and an array of oak woodlands, serpentine chaparral, closed-cone pine and cypress groves, and sparkling wildflower displays. It also holds some of the best examples of California sycamore (*Platanus racemosa*) riparian woodland in the state. At the southern terminus of the Diablo Range, the Temblor Range continues to the east-west trending Tehachapi Mountains where the eastern mountain ranges of the Central Coast Ranges end. To the west of the Temblor Range is the Carrizo Plain, which is bounded to the west by the La Panza Range and the coastal Santa Lucia Range. These southern mountains and valleys are distinctively different than those to the north. The Carrizo Plain and the Temblor Range have a semi-desert climate. This landscape includes a dry, alkali lake (Soda Lake) and slot canyons, and it shares a number of plant species with the Mojave Desert.

Within this broad region is an expansive landscape of wilderness within

a matrix of coastal and interior oak woodlands, chaparral, grasslands, and coastal scrub. In many portions of this region the pace of human life has not changed appreciably for 100 years. Most inhabitants live on ranches, vineyards, or in occasional small towns in the valleys, even though large cities and valuable agricultural lands exist to the west and east. Much of the core portion of the Diablo Range is wilderness. Large watersheds are essentially roadless and many streams are without dams or bridges.

The region's climate is markedly altered by the coastal Santa Cruz Mountains and Santa Lucia Mountains. These ranges intercept much of the moisture from Pacific storms and create a rain shadow that deprives most of the inner Coast Ranges of abundant precipitation. Drawing in that moisture, Cone Peak in the Santa Lucias is notable for its groves of canyon live oak (*Quercus*

chrysolepis), sugar pine (*Pinus lambertiana*), and Santa Lucia fir (*Abies bracteata*). While it is not unusual for many parts of the Santa Cruz and Santa Lucia mountains to receive precipitation in excess of 700 mm annually, average annual precipitation in the inner valleys and ranges is typically around 400 mm. Even the highest mountain peaks (Mount Diablo, Mount Hamilton, and San Benito Mountain) in the Diablo Range seldom exceed 600 mm average annual precipitation. Not surprisingly, the rain shadow creates hot and arid conditions over most of the region.

Except in the higher mountains, where average annual temperatures range from 20–25°C, the majority of the region averages more than 30°C. Nevertheless, two coastal openings enable summer fog to penetrate into the interior of the Central Coast Ranges. The fertile Salinas Valley via the Monterey Bay and the Santa Clara Valley via the Golden Gate and San Francisco Bay (on inward to the western Sacramento-San Joaquin Delta) are bathed by fog that permeates into the adjacent mountains. The entire region sits almost entirely atop what is known geologically as the Salinian block, which is bounded on the east by the San Andreas Fault and on the west by the Nacimiento Fault. A number of lesser known faults are arranged parallel to these major ones and are in part responsible for the parallel pattern of ridges and valleys characteristic of the region. At its core, the Salinian block is primarily granitic, while the

Top: The Santa Lucia Mountains intercept moisture and contain unique groves of hardwoods and conifers including canyon live oak, sugar pine, and Santa Lucia fir.
JULIE M. EVENS

Left: Mature valley oaks growing in deep alluvial soils near massive, uplifted sandstone outcrops along the San Antonio River Valley to Santa Lucia Memorial Park.
TODD KEELER-WOLF

rocks on either side of these major faults are largely Franciscan sedimentary and metamorphic. In the Diablo Range, however, Franciscan rocks are relatively common and mix with younger sedimentary rocks. The Franciscan series is important to vegetation composition because serpentinite parent material, and the soils that weather from it, are commonly interspersed with these rocks. Serpentinite lacks nitrogen, phosphorous, and calcium that are vital to plant growth, and contains many toxic heavy metals such as magnesium, lead, chromium, and mercury. Wherever serpentinite is scattered about the inner Coast Ranges, rare and endemic plant species are often present. On Coyote Ridge to the east of the Santa Clara Valley, serpentine supports the endemic coyote ceanothus (*Ceanothus ferrisiae*). Here serpentine grasslands and wildflower fields support the last remaining native populations of the Bay Checkerspot butterfly (*Euphydryas editha bayensis*). This butterfly feeds on the diminutive dwarf plantain (*Plantago erecta*), found in abundance on serpentine soils, which repel non-native aggressive grasses. On San Benito Mountain (1597 m) shearing of the serpentine has exposed heavy metals and asbestos creating bizarre, natural barrens which support unusually open forests of isolated montane conifers.

Blue Oak and Foothill Pine Woodlands

Blue oak (*Quercus douglasii*) and foothill pine (*Pinus sabiniana*) often combine to form woodlands, which are collectively part of the blue oak alliance. These woodlands cover much of the expansive valleys and gently rolling foothills of the Central Coast Ranges. The blue cast of the oak canopies is often punctuated by the ghostly appearance of tall, scattered foothill pines, which taken together lend a distinctive character to the woodlands, and to the greater landscape. Together these two species provide welcome visual relief from what otherwise would be an endless expanse of annual grasslands and other herbaceous vegetation.

Top: Serpentine grasslands with wildflowers along Coyote Ridge in the foreground and oak woodlands along Mount Hamilton Range in the background. Julie M. Evens

Above: Bay Checkerspot butterfly. Stuart Weiss

Top: Blue oak woodlands along the rolling slopes at Tesla Park in southeast Alameda County.
Pete Veilleux

Middle: Blue oak woodland punctuated with tall scattered foothill pines in inner Santa Barbara County.

Left: Shooting stars (Dodecatheon clevelandii) *in the understory of blue oak woodland at Chimineas Ranch.*
Todd Keeler-Wolf photos

At first glance, these woodlands appear to be lacking in variety of plant species, especially when summer turns their understory to a uniform golden hue. But they are surprisingly rich in native plants, especially where the trees grow denser and the cumulative shade of the oak canopies crowds out the open grasslands. The understory's uniform appearance is the result of near-complete domination of a handful of introduced Mediterranean grasses. But before these annual grasses reach their peak in late spring, a rich array of low-statured native annual herbs and perennial forbs make a brief appearance. Where oak woodlands become forests (densities > 620/ha), it is possible to find as many as 60 native plant species in 400 m^2 (Borchert et al. 1993). Moreover, many of these natives owe their existence to the oaks which ameliorate soil, water, and light conditions beneath the canopy.

In this sunny Mediterranean climate, evaporation easily surpasses annual precipitation and creates an increasingly severe soil water shortage as the summer progresses. Annuals avoid this hostile climate by growing through the winter and reproducing in early spring. In contrast to these ephemerals, blue oak and foothill pine are in full leaf throughout summer and well into the fall and therefore must find ways of overcoming the persistent heat and moisture stress.

Over the millennia, blue oak has evolved structural and functional traits to cope with this seasonally arid climate. For example, it reduces water use by dropping its leaves in late summer and fall, and by maintaining a relatively open canopy in spring and summer. Compared to oaks living in moister environments, like coast live oak (*Quercus agrifolia*), it has smaller leaves that stay 2–3° C cooler than oaks with larger leaves. And functionally, photosynthesis continues during the lowest periods of soil water in late summer and early fall because roots tap into deep-water sources including cracks in the bedrock. In fact, blue oak is capable of "hydraulic lift," meaning it redistributes water from wet soils at depth to drier soils near its roots (Baldocchi and Xu 2007).

Valley Oak Woodlands and Savannas

Among California's endemic oaks, none is as stately as valley oak (*Quercus lobata*). Its twisted, spreading branches and massive trunk create a distinctive silhouette recognizable from a considerable distance. Like blue oak, valley oak is deciduous, but unlike blue oak, valley oak is not so drought tolerant. In order to persist in this dry climate, its roots must grow deep enough to reach the predictable soil moisture of valley-bottom water tables. In the fall its long, teardrop-shaped acorns litter the ground attracting mule deer (*Odocoileus hemionus*) and many other wildlife species. Acorns in the canopy are harvested by acorn woodpeckers (*Melanerpes formicivorus*) and western scrub jays (*Aphelocoma californica*).

Unfortunately, it is becoming increasingly difficult to find extensive areas where valley oak is numerous and dominant. Like California's riparian areas, in general, valley oak woodlands have undergone a dramatic areal reduction, principally because they grow in the deep, alluvial valley bottom soils that are highly suited to agriculture. Recent studies in the Santa Clara Valley suggest that valley oaks have been reduced by as much as 90% since the arrival of the early settlers (Grossinger et al. 2007). There are, however, a few remaining locations that give a flavor of how these woodlands looked 200 years ago. On the road from the Nacimiento Valley to the Big Sur coast that crosses Fort Hunter Liggett, the valley oaks are amazing in both size and form. Fort Hunter Liggett and the surrounding valleys on Los Padres National Forest, such as near "The Indians," have some impressive stands of valley oak woodland with a diverse understory of native herbaceous species such as curly

Above: The twisted, spreading branches of a valley oak on Mount Hamilton create its highly recognizable open canopy.
Jeff Bisbee

Below: An individual western scrub jay can bury thousands of acorns in the soil and retrieve most over the course of the following year.
Todd Keeler-Wolf

blue grass (*Poa secunda*), deer grass (*Muhlenbergia rigens*), nodding needle grass (*Nassella cernua*), Turkish rugging (*Chorizanthe staticoides*), popcorn flower (*Plagiobothrys nothofulvus*), goldfields (*Lasthenia californica*, *L. gracilis* and/or *L. leptalea*), and many other native annual and perennial species.

Shrubby Oaks and California Junipers

Inland and eastward from the Salinas Valley, the inner Central Coast Ranges experience a semi-desert climate. Summer temperatures grow increasingly hot, and summer-fall drought becomes ever more extreme. On the dry slopes facing the San Joaquin Valley grow stands of semi-desert vegetation that the casual traveler might easily miss. On slopes and ridges overlooking the distant agricultural lands of the San Joaquin Valley, Tucker oak (*Quercus john-tuckeri*) chaparral grows on north-facing slopes where it frequently co-mingles with shrubby California junipers (*Juniperus californica*). Tucker oak is a short shrubby tree that regularly hybridizes with blue oak where their ranges overlap.

On the hot, steep rocky slopes grow expansive stands of California buckwheat (*Eriogonum fasciculatum*) scrub bristling with the spiny rosettes of chaparral yucca (*Hesperoyucca whipplei*). Indeed, many of the shrubs in this dry, rocky setting are widespread in the mountains of the Mojave and Great Basin deserts,

Top: Valley oak woodland with a rich native plant understory near "The Indians" in the upper San Antonio River valley, adjacent to the Santa Lucia Mountains.

Middle: California junipers with an understory of purple sage at Chimineas Ranch, eastern San Luis Obispo County.
Todd Keeler-Wolf photos

Left: California buckwheat with chaparral yucca.
Gina Radieve

and include species such as California joint fir (*Ephedra californica*), narrow-leaf goldenbush (*Ericameria linearifolia*), rubber rabbitbrush (*Ericameria nauseosa*), and big sagebrush (*Artemisia tridentata*). Combinations of these same species may be found on the western edge of the Mojave Desert on the eastern slopes of the Tehachapi Mountains, southern Sierra Nevada, and the northern slopes of the San Gabriel Mountains.

In the sandy soils on the low slopes immediately above the San Joaquin Valley grow scattered and isolated groves of the unusual Alvord oak (*Quercus × alvordiana*), a hybrid between Tucker and blue oak. The most spectacular stands of this hybrid occupy sandy ridges near Cantua Creek north of New Idria in western Fresno and San Benito counties. These oaks typically inhabit coarse-textured soils where annual grasslands are sparse. Because this discontinuous grass cover does not readily carry fire, the oaks are spared tree-killing fires that would normally eliminate them.

Coast Live Oak Forests and Woodlands

As its name implies, coast live oak seldom grows more than 100 km from the Pacific coast. It forms conspicuous dense evergreen forests near the coast that are very different from the open blue and valley oak woodlands that cover the interior valleys. Progressing from the coast to the interior of the Central Coast Ranges, coast live oak is more likely to mingle with valley and blue oaks in open woodlands rather than form dense forests. And before it disappears from the landscape altogether, it retreats to riparian areas and canyons where it finds enough water to endure the scorching summers.

Unlike the leaves of blue and valley oaks, coast live oak leaves are ever-

Above: California joint fir shrubland with goldfields along the Elkhorn Plain.
Julie M. Evens

Below: Acorn woodpecker storing oak acorns in a granary tree.
John Sterling

Above: Shrubby Tucker oak in Potrero Canyon in the Temblor Range.
Julie M. Evens

Right: Alvord oaks with narrowleaf goldenbush in the Caliente Range, San Luis Obispo County. Evergreen Tucker oak and deciduous blue oak often co-occur at the interface between chaparral and oak woodland.
Todd Keeler-Wolf

green and leathery. In forest settings, dead leaves accumulate beneath the dense canopies often forming a deep litter layer that decomposes slowly. As a consequence, most blue oak understory herbaceous natives are absent from dense coast live oak forests and are replaced by more shade-tolerant species such as hummingbird sage (*Salvia spathacea*) and fiesta flower (*Pholistoma auritum* and *P. membranaceum*). On the other hand, a wider variety of understory shrub species grow with coast live oak than with blue oak. In the moist coastal forests, understory shrubs include blackberry (*Rubus* spp.), poison oak (*Toxicodendron diversilobum*), hazelnut (*Corylus cornuta*), oceanspray (*Holodiscus discolor*), California coffee berry (*Frangula californica*), snowberry (*Symphoricarpos* spp.), and toyon (*Heteromeles arbutifolia*). The unusual fire ecology of coast live oak explains in part its ability to live with continuous understories of highly inflammable shrubs like chamise (*Adenostoma fasciculatum*) and the sages (*Salvia* spp.). Unlike other California oaks, coast live oak has a thick, corky bark that protects its fragile inner cambium from lethal temperatures of even the most intense fires (Plumb and Gomez 1983). After a wildfire, it is not uncommon to find coast live oaks with all the leaves burned off and the trunk and branches deeply charred. By spring, up to 90% of the immolated, apparently dead trees come back to life primarily because the thick bark protects dormant buds scattered along the branches. Buds sprout and shoots poke through the charred bark or appear around the base of the trunk.

Herbaceous Vegetation

In similar dry settings grow desert needlegrass (*Achnatherum speciosum*), the beautiful endemic shrub yellow mock aster (*Eastwoodia elegans*) with its large, yellow, globular flowering heads. Even typical desert plants such as Indian rice grass (*Achnatherum hymenoides*) grow here. Up and down the driest hills of the inner Coast Ranges are numerous plant species and communities that reflect the near-desert environment of this part of California. Under certain climate change scenarios, one could contemplate the expansion of this patchy semi-desert vegetation as possible in the coming decades.

In addition to the desert grasses and

Top: Dense coast live oak woodland on shallow soils interspersed with grassland and chaparral in Carmel Valley, Monterey County.

Above: Coast live oak woodland with lush understory of ferns, blackberry, and hazelnut in the Oakland Hills.

Todd Keeler-Wolf photos

shrubs, a wealth of annual herbs can cover neutral and south-facing slopes and valleys in good rainfall years. With the right timing of precipitation, vast areas are painted with brilliant swatches of yellow, orange, blue, and purple. Species creating this riot of color include the endemic Californian genus *Monolopia*, the orange-hued genera *Mentzelia* and *Amsinckia*, and the lavender, blue, and purple species of genera such as *Phacelia*, *Salvia*, *Lupinus*, *Gilia*, and *Nemophila*. The heavy clay slopes of inner stretches of the central Coast Ranges from eastern Alameda to eastern San Luis Obispo counties support some of the finest wildflower displays still left in California. Such wildflower displays have vanished from much of more coastal and southern California but are still significant in this region.

Coulter and Foothill Pine Woodlands

Coulter pine (*Pinus coulteri*) and foothill pine are the most common conifers in the Central Coast Ranges. These pines are taxonomically closely related and highly drought-tolerant, although detailed studies of their physiology have been conspicuously few. Foothill pine woodlands occur on hot, dry valley floors with blue oak, valley oak, and coast live oak often co-dominating. In contrast, Coulter pine woodlands replace foothill pine woodlands at higher elevations; the distributions of the two species

Top: Desert needlegrass on dry, sandy soils.
CNPS

Middle: Desert-like sand ridges on the Los Angeles La Panza Range with mixes of desert dandelion (Malacothrix glabrata) *and thistle sage* (Salvia carduacea).
Todd Keeler-Wolf

Bottom: Lupine and California poppy (Eschscholzia californica) *display along the Carrizo Plain.*
Robert Wick

Above: Some of the best displays of springtime wildflowers occur along Highway 58 and Shell Creek Canyon south of Shandon, an area where fossilized shells are found in the exposed strata of the bordering hills.
Julie M. Evens

*Left: Colorful wildflower display consisting of California poppy, tidy-tips (*Layia platyglossa*) and lupine at the Carrizo Plain.*
Robert Wick

Above: Coulter pine over scrub oak at the top of San Benito Mountain.
TODD KEELER-WOLF

Opposite top: Post-fire regeneration of Coulter pine and wartleaf ceanothus in the Santa Lucia Mountains.
JEFF BISBEE

typically do not overlap. Genetic studies have shown that the two species differ markedly in how they expanded over the Central Coast Ranges and California in the past. Coulter pine has its origins in the Laguna Mountains near San Diego and over time has spread north along two routes: one along the coast and the other inland through the valleys of the Coast Ranges (Ledig 2000). Although its seeds are winged, they are heavy, with poor long-distance dispersal capabilities. Apparently, local dispersal and regeneration depends more on small mammals and jays than on wind. Given its relatively poor seed dispersal, it is surprising that Coulter pine was able to advance so far northward in a stepping-stone manner, with individual populations sometimes separated from others by seemingly insurmountable distances. The colonizing history of foothill pine is more diffuse than that of Coulter pine.

While Coulter and foothill pines share many morphological and ecological characteristics, they differ markedly in fire ecology. Although both species have a relatively thick bark that protects them from fire, both are highly sensitive to crown scorching and—compared to pines like Jeffrey (*Pinus jeffreyi*) and ponderosa (*P. ponderosa*)—fire carries easily into their long-needled, dense canopies. Foothill pine apparently tolerates little more than the low-intensity surface fires that carry through the annual grasses. In fact, any amount of understory shrub cover almost guarantees fire-caused tree death.

Coulter pine has a very different response to moderate and high intensity fires. In the La Panza Range, for example, it grows over a continuous understory of chaparral shrubs and is immediately killed by fire. However, Coulter pine is serotinous. The heavy, well-armed cones first appear in 10-year-old trees and grow attached to the trunk since they are too heavy to be supported by the branches. As the tree ages, the accumulating cones

do not open and drop but remain closed and attached to the tree. In effect, the tree stores viable seeds in these closed cones waiting for the inevitable fire. When the next high-intensity fire rages through the understory chaparral, it easily kills overstory trees. At the same time, heat melts the resinous seal of the cones, releasing the seeds *en masse*. Seeds litter the blackened soil and are quickly collected by rodents who eat many, but bury and forget the rest (Borchert 1985). Seeds not re-located germinate into conspicuous seedling clusters. Seedlings grow quickly and thin out over time to become saplings that once again produce closed cones that will open in the next cycle of fire.

Higher Elevation Relict Trees and Unusual Soils

Deep in the inner Coast Ranges of San Benito County lies the unearthly Mount San Benito–Clear Creek terrain with its bluish-green barren slopes, creeks that run well into the summer, and jumbled landscape of cliffs and landslides. From this extensive serpentinite intrusion, miners once extracted mercury, asbestos, and other rare minerals, and left huge scars dotting the landscape. Among the many unusual plant species in this terrain are isolated stands of Jeffrey pine and incense cedar (*Calocedrus decurrens*), conifers usually encountered in mid to high elevations of the Sierra Nevada, Tehachapis, and Transverse Ranges of southern California. The woodlands in the Jeffrey pine alliance are over 125 km from similar stands to the south. Accompanying these conifer stands are several shrubland alliances which are adapted to the rigors of serpentine soils. The leather oak (*Quercus durata*) alliance is the main shrubland on rocky but sheltered slopes, while the bigberry manzanita (*Arctostaphylos glauca*) alliance forms stands on the rocky ridgetops. Along

Above: The open stands of Jeffrey pine, Coulter pine, and incense cedar, growing on serpentine on San Benito Mountain, are isolated from the nearest mixed conifer stands over 120 miles to the south.
Todd Keeler-Wolf

Top: Jeffrey pine and incense cedar rise on the higher elevation slopes of San Benito Mountain.

Above: Rayless tidy-tips is endemic to serpentine on San Benito Mountain.
Todd Keeler-Wolf photos

Right: Bigberry manzanita and foothill pine on a south-facing, serpentine slope in Clear Creek.
Julie M. Evens

the rocky streambeds, the Brewer willow (*Salix breweri*) alliance forms patchy riparian shrublands along with hoary coffee berry (*Frangula californica* ssp. *tomentella*) and endemic herbs such as Van Houtte's columbine (*Aquilegia eximia*). Rare endemic herbs such as rayless tidy-tips (*Layia discoidea*) and disjunct isolated populations of plants such as pointleaf manzanita (*Arctostaphylos pungens*), occur on the rocky upland slopes (Evens et al. 2006). All survive here because the nutrient imbalance of the serpentinite soils reduces competition with regionally common species that are better adapted to non-serpentine soils in this dry interior climate (Alexander et al. 2006).

Chaparral and Coastal Scrub

Shrublands of the region include evergreen thick-leaved chaparral and drought-deciduous soft-leaved coastal scrub, especially along rocky ridges and steep slopes with thin soil. Moist north-facing slopes and draws are often laden with ecologically similar stands of tall chaparral vegetation including the holly leaf cherry (*Prunus ilicifolia*), toyon, birch leaf mountain mahogany (*Cercocarpus montanus*), scrub oak (*Quercus berberidifolia*), and Tucker oak (*Q. john-tuckeri*) chaparral alliances; all with slightly different ecological niches, which sort out by slope position and orientation. These shrublands are interspersed with trailing perennial lianas and vines such as pipestem clematis (*Clematis lasiantha*), pink honeysuckle (*Lonicera hispidula*), and large fruited man-root (*Marah macrocarpus*). On sunny, drier slopes, shrub stands shift to different dry-adapted chaparral alliances including chamise and wedge leaf ceanothus (*Ceanothus cuneatus*). On warm sunny slopes with finer soil texture the coastal scrub alliances such as purple sage (*Salvia leucophylla*), California sagebrush (*Artemisia californica*), black sage (*Salvia mellifera*), and California

Top: Brewer willow in a sepentinite creekbed near San Benito Mountain.
Jeanne E. Taylor

Above: Birch leaf mountain mahogany with seeds giving a feathery look.
Julie M. Evens

Top left: Chamise growing with bigberry manzanita on the drier slopes in Pinnacles National Park.

Top right: Holly leaf cherry shrubland above a blue oak–pine woodland on a moist slope in Pinnacles National Park.
Julie M. Evens photos

Above: Wartleaf ceanothus along the post-burn slopes, five years after the 2008 Indians Fire in the Santa Lucia Mountains.

Center: Beautiful fruits of of the endemic Gabilan manzanita.
Todd Keeler-Wolf photos

buckwheat (*Eriogonum fasciculatum*) frequently dominate.

Sedimentary slopes closer to the coast are distinguished by a diversity of other chaparral alliances including Eastwood's manzanita (*Arctostaphylos glandulosa*), hairy leaf ceanothus (*C. oliganthus*), and wart leaf ceanothus (*C. papillosus*). In the Gabilan Range, such alliances are punctuated by the recently discovered, rare and beautiful Gabilan manzanita (*A. gabilanensis*).

Conservation of the Landscape

Most areas of the Central Coast Ranges have seen little of the expansion of cities and agriculture that dominate

the more coastal or more inland valleys of Central California. Nevertheless, in the past three decades, the area has been subjected to suburban and vineyard expansions, renewable energy development, and power corridor and reservoir construction. Some public lands, such as BLM's Clear Creek Management Area, were originally established for multiple-use including OHV activity that is detrimental to some of the local vegetation. Ironically, this area is currently closed due primarily to public safety concerns over the erosion and the carcinogenic dust produced when some serpentine outcrops are disturbed by this form of recreation. To provide water for agriculture and other human consumption, water storage areas like the San Luis Reservoir have inundated hundreds of hectares of riparian California sycamore woodlands. Unlike the higher elevations of the mountains, where extensive and interconnected federal Forest Service and Park Service wilderness areas have been established, other lower elevation parks and preserves comprise the largest existing open spaces in the Central Coast Ranges and tend to be private ranches. Many of the ranches have persisted for long periods. They should be considered viable parts of the natural landscape and vital to the overall high ecological integrity of the region.

Among the valuable and interesting conservation lands are several large parcels in the Diablo Range and the Gabilan Range, in which The Nature Conservancy has acquired easements. These connect to extensive public or publically managed lands such as the Ohlone Wilderness (managed by East Bay Regional Parks) and Henry W. Coe State Park. Many creative conservation

Top: The core of the Diablo Range north of Mount Hamilton in the Ohlone Wilderness in Alameda County.
Todd Keeler-Wolf

*Above: Tule elk (*Cervus canadensis nannodes*) rely on large intact natural landscapes, such as at the Wind Wolves Preserve.*
Jennifer Buck-Diaz

Above: Giant kangaroo rat burrow and precinct with Great Valley phacelia (Phacelia ciliata) *growing on it and with goldfields in the background.*
Todd Keeler-Wolf

Inset: Giant kangaroo rat and burrow on the Carrizo Plain.
Robert Wick

Opposite: The Carrizo Plain National Monument is the largest of three core recovery areas identified by the U.S. Fish and Wildlife Service for preserving upland species of the San Joaquin Valley. It supports thirteen state and federally listed plant and animal species amongst an incredible diversity of forbs, grasses, and shrubs.
Robert Wick

efforts have been launched in the area, including one of the shining stars, the Carrizo Plain National Monument, which is an exemplary conservation and management collaboration among The Nature Conservancy, Bureau of Land Management, and California Department of Fish and Wildlife. When added to other state and federal lands managed for wildlife values, it totals more than 100,000 ha.

Giant Kangaroo Rats

Many vertebrate species inhabit the Carrizo Plain, but the giant kangaroo rat (*Dipodomys ingens*) plays an outsized role in shaping composition, structure and function of plant and animal communities of these semi-arid grasslands. Giant kangaroo rats are ecosystem engineers because, although relatively small in size, they change biotic and abiotic characteristics of this grassland habitat in many unexpected ways.

Their burrows tend to be evenly spaced over the flat terrain. Over many years, they dig new burrows as well as enlarge and maintain existing ones. Pushing soil to the surface creates a low mound called a "precinct" that radiates from the burrow entrance. Precincts vary in size from 1.8 to 7.7 m in diameter and can cover up to 32% of the grassland area. Since kangaroo rats are primarily seed-eaters, they clip annuals growing on their precincts and allow the seed heads to ripen. Most of the seeds are carried below ground where the kangaroo rats store them for future use. Seeds are cached above ground in small pits scattered around the precinct. Invariably, seeds in some pits are forgotten and later germinate to become clusters of annuals that will serve as a future food source.

Because kangaroo rats prefer large, non-native seeds such as red brome (*Bromus rubens*) and redstem filaree (*Erodium cicutarium*), these species dominate the precincts at the expense of natives. Paradoxically, this small-scale precinct dominance by non-natives does not extend to larger scales. At the landscape scale, kangaroo rat activities actually increase plant species diversity, resulting in more natives than non-natives (Prugh and Brashares 2012).

Giant kangaroo rats also influence vertebrate and invertebrate species communities. State-threatened San Joaquin antelope squirrels (*Ammospermophilus nelsoni*) use the burrows to escape predators as well as hot summer temperatures. Antelope squirrel numbers are higher where kangaroo rats are present than where they are absent—as are lizard species that also use burrows for escape and thermoregulation. Invertebrates such as orthopterans (grasshoppers, crickets, etc.) and beetles occur in higher numbers where kangaroo rats dig burrows, perhaps benefiting from their stored plant food.

Mark Borchert
Todd Keeler-Wolf

CHAPTER TEN

Central California Coast

Stretching along the coast from the San Francisco Bay south to Point Conception, west of the Santa Clara, Salinas, and Santa Ynez Valleys, the Central Coast is both a botanical melting pot and a center of endemism. Its vegetation includes a wide variety of species which reach their northern, southern or western limits here. However, it also contains an impressive list of unique plant species, as demonstrated by more than 30 species endemic to Monterey County alone (Matthews 2006). This diversity is a product of local climatic, geologic, edaphic and floristic diversity. As with the array of species found in the region, the vegetation of the Central Coast can be wide-ranging or narrowly endemic, unique to the area or characteristic of much of California.

California's Central Coast and adjacent Central Coast Ranges are sandwiched between the Pacific Ocean and the Great Valley. This chapter showcases the Outer Coast Ranges, west of the north-south trending valleys between the Inner and Outer Coast Ranges, while the chapter on the Central Coast Ranges focuses on the Inner Coast Ranges and adjacent valleys. At the edge of the continent, the Pacific Ocean delimits the Central Coast's western boundary and is the predominant influence on its climate. Storms originating over the Pacific Ocean drop much of their moisture over the outer Coast Range, due to the cooling effects of orographic lift and resulting condensation, effectively leaving the Diablo Range, the inner Coast Range, and the Great Valley in

Above: Coastal grasslands and chaparral reflect the region's variation in vegetation in Big Sur, Monterey County.
PETE VEILLEUX

Opposite: The diversity of Central Coast plant communities reflects the area's myriad geologic, edaphic, cultural, and local climatic conditions.
STEVE PERDUE

Top: The Pacific Ocean is a primary influence on the region's climate, generating the characteristic summer fog. Bixby Bridge in Big Sur, Monterey County.
JEFF BISBEE

Above: The array of vegetation across the landscape often indicates the complex underlying geology of the region. Santa Lucia Mountains, Monterey County.
PETE VEILLEUX

a rain shadow. Parts of the Santa Lucia Mountains average up to 165 cm of precipitation per year, while in their shadow the adjacent Salinas Valley receives only 28 cm. Additionally, the upwelling of colder subsurface ocean waters produces the region's characteristic coastal fog and generally acts to buffer the extremes of California's Mediterranean climate along the coast, rendering a much gentler version of the typical inland climate of hot, dry summers and cold, wet winters. Coastal fog occurs primarily in the summer months, and is most persistent where there are coastal connections to large interior valleys. Average annual precipitation ranges from about 40–50 cm at the coast and lower-elevation valleys to over 130 cm on the higher slopes of the coastal mountains. The area has a very moderate climate, with average temperatures 10–30°C in summer and 0–20°C in winter. Although frost is common in winter, especially in the coastal valleys, snow is a rare event.

The geology of the area—as reflected in the landforms and soils, as well as their mineral composition—is an important architectural agent of the vegetation and its arrangement across the landscape. The Central Coast Ranges are primarily composed of the Franciscan assemblage, a contorted mix of materials of various origins, including sediments deposited on the ocean floor into which lava or other igneous rock were forced in molten condition, in some places altering, or metamorphosing, the raw geologic materials into new materials. The resulting mix has been faulted, folded and uplifted in a seemingly chaotic manner along the margin of the North American Plate to form the Coast Ranges, although a view from 200 miles up reveals the northwest-to-southeast trending ridges and valleys resulting from concurrent tectonic movement along the San Andreas Fault and associated right-lateral faults.

Once uplifted, subsequent erosion of softer material has occurred, and in places deposition of alluvial fans and filling in of valley bottoms. Notable outer Coast Range peaks—including Loma Prieta in the Santa Cruz Mountains, Marble Cone, Ventana Cone and Cone Peak in the Santa Lucia Mountains, and Cerro Alto and Morro Rock/the Seven Sisters in San Luis Obispo County—are remnants of ancient volcanic and metamorphosed sedimentary deposits, uplifted in this seismically active area over tens of millions of years. Along the immediate coast examples of the Franciscan assemblage, as well as uplifted marine

sediments, form terraces above the shore. More modern (Pleistocene and Holocene) features, such as dunes and alluvial terraces, formed after primary uplifting.

The outer Coast Ranges, particularly the Santa Cruz Mountains and the Santa Lucia Range, are acutely uplifted and characterized by short, steep drainages connecting directly west to the ocean. These rugged ranges benefit the most from the effects of orographic lift, squeezing out much of the area's precipitation before it gets very far inland; additionally, summer fog acts to supplement effective precipitation through "fog drip." Along streams and on moist slopes in these ranges can be found the southern extent of a number of forest alliances which are more common to the north. These include forest alliances featuring redwood (*Sequoia sempervirens*), tanoak (*Notholithocarpus densiflorus*), madrone (*Arbutus menziesii*) and Douglas fir (*Pseudotsuga menziesii*) as dominants

Top: Sediments deposited under the ocean and then uplifted can be seen overlaying igneous rock adjacent to this cove. North Coast of San Luis Obispo County.
Deborah Hillyard

Inset: Farewell-to-spring (Clarkia rubicunda) *reaches its southern extent in San Luis Obispo County, here on the bluffs of the north coast.*
Dave Hacker

Above: Redwood and other forest alliances on the moist, north-facing slopes and drainages contrast the scrub and prairie types on the opposite, drier slopes in Big Sur.
John Stuart

Top: The inland portion of the Central Coast, much more arid, hosts abundant annual wildflowers interspersed into the mix of annual grassland and blue oak woodland dotted with foothill pine. Shell Creek, San Luis Obispo County.
DEBORAH HILLYARD

Above: Formerly coastal scrub, riparian and oak woodland, and prairie, large areas of native vegetation have been converted to agriculture, adjacent to the Salinas River.
ANDREA PICKART

or co-dominants, as described in more detail in the North Coast section.

The significant climatic, geologic, and topographic variability of this region overall, along with its medial location, is reflected in its floristic diversity. The immediate Central Coast, delimited from the interior ranges by the larger valleys such as the Salinas Valley and Santa Ynez Valley, is a matrix of common and characteristic plant communities rife with pockets of interesting and unusual assemblages. With the exception of the Santa Cruz and Santa Lucia Mountains, the preponderance of the Coast Ranges is less rugged and drier, and supports a distinctive mosaic of woodlands, chaparral, scrublands and grasslands, as described in the Central California Coast Ranges chapter. For example, California juniper (*Juniperus californica*) and blue oak (*Quercus douglasii*) woodlands dot the hills and valleys of the drier interior, while many endemic cypress (*Hesperocyparis* spp.), pines (*Pinus* spp.), and coast live oak (*Quercus agrifolia*) forests and woodlands drape the slopes nearest the coast. Not unexpected due to the area's central location, the entire region is unique state wide in supporting a mixed-oak forest which contains three or more co-dominant species of the oaks (*Quercus* spp.)

The pattern and character of vegetation across the Central Coast landscape have changed since western explorers first set eyes on it in the 1700s. Due to increasing human population and attendant activities, a vast majority of the deep, rich soils of the coastal terraces and broad floodplains deposited by the region's biggest rivers have been converted to urban areas or intensive agriculture, and the remaining grasslands and woodlands that did not succumb to wholesale conversion have been altered by over 200 years of grazing by domestic livestock and the introduction of invasive annual grasses and forbs. In the more remote wildlands, the role of fire has changed from a regular and important Native American landscape management tool to an irregular and serious threat to human and natural resources. Despite these changes, the Central Coast contains some of the most quintessential Californian vegetation patterns synonymous with our visions of "old" California's mission and ranching history. Its wildflower-washed open oak woodlands, chaparral-clad plunging

ridges, deeply-incised coastal canyons, bluffs perched over the scenic shoreline, and coastal wetlands all stand as icons of California landscapes.

Coastal Estuaries

Estuarine areas are characterized by the mixing of freshwater coming from inland streams with salt water from the Pacific Ocean. Estuaries buffer streams from the relentless action of the surf, act to capture sediment and pollutants originating in the uplands, and serve as a nursery for much of the near-shore marine biota. The pattern of vegetation in estuaries is indicative of the array of physical and chemical stressors, such as flooding and salinity, which characterize this important transitional habitat. Global climate change and its anticipated rise in sea-level could result in the relatively rapid loss of estuarine habitats through inundation. As such, it is considered a major threat to this ecosystem.

The most extensive coastal wetland in California (and indeed, the entire west coast of North and South America) is the San Francisco Bay Estuary, extending from the Golden Gate inland to the junction of the Sacramento and San Joaquin Rivers. Like much of California's wetlands, the San Francisco Estuary has been greatly altered and reduced in overall extent over the last 200 years. Elkhorn Slough and Morro Bay are also prominent coastal estuaries in the region. Coastal wetlands may still be found at the mouths of many coastal streams, including some of the larger streams along the coast, such as Pescadero Creek, Salinas

Above: Even though estuaries are inundated twice each day by high tides, the habitats of the virtually flat marsh of Morro Bay estuary, as with all coastal estuaries in California, are extremely vulnerable to any rise in sea level.

Inset: Found in estuaries only in a narrow band typically near the upper reaches of pickleweed marsh, salt marsh bird's beak is a species which has declined throughout its range as wetlands have been lost.
Deborah Hillyard photos

Top: Pickleweed mat with reddish fall color, Martin Luther King, Jr. Regional Shoreline Park.
PETE VEILLEUX

Above: California cordgrass marsh in San Francisco Bay.
JOHN O. SAWYER

River, Carmel River, Big Sur River, Santa Maria River, San Antonio Creek, and Santa Ynez River.

Pickleweed (*Sarcocornia pacifica*) mats alliance, often in combination with salt grass (*Distichlis spicata*) flats alliance in the higher reaches, is the most common vegetation type of the larger estuaries, developing on the flat marsh plain. Although normally emergent, pickleweed mats are typically inundated twice a day during highest high tide events. Other associated marsh species include alkali heath (*Frankenia salina*), seaside arrowgrass (*Triglochin maritima*), gum plant (*Grindelia stricta*) and marsh jaumea (*Jaumea carnosa*). In some years, salt marsh dodder (*Cuscuta salina*) forms a conspicuous mat over the marsh vegetation. Salt marsh dodder is a parasitic herbaceous flowering vine with a very

notable bright orange color, which denotes a lack of chlorophyll resulting in the inability to photosynthesize.

Closer to the open water, California cordgrass (*Spartina foliosa*) marsh alliance characterizes the lower marsh settings in San Francisco and Monterey Bays. California cordgrass marsh has been greatly altered by the introduction of several non-native cordgrass species (for example, *S. alterniflora*, which hybridizes with California cordgrass, and *S. densiflora*), which can tolerate a wider range of inundation than the native species; they have invaded mud flats and adjacent marshes in San Francisco Bay, and threaten to eliminate California cordgrass marshes in these areas.

At the upper reaches of most coastal estuaries, in more protected areas with ample fresh water, the vegetation transitions to that which is characteristic of other coastal freshwater wetland vegetation: willow thickets, typically dominated by arroyo willow (*Salix lasiolepis*) alliance; cattail (*Typha* [*angustifolia, domingensis, latifolia*]) marshes; and bulrush marshes characterized by California bulrush (*Schoenoplectus californicus*) and/or hardstem bulrush (*S. acutus*).

California sea-blite (*Suaeda californica*) was once found at the upper margins of estuaries, around the San Francisco Bay south to at least Estero Bay in San Luis Obispo County, and, as indicated by plant remains in adobe bricks, perhaps as far north as the Petaluma River. As coastal marsh habitat has been lost, this species now survives naturally only adjacent to the upper tidal edges of Morro Bay, although it has been reintroduced in several locations at Golden Gate National

Protected from the surf and tides, and farther from the effects of salt water, willow thickets flourish in the dunes at Oso Flaco Lake, San Luis Obispo County. Deborah Hillyard

Above: Dominated by California oat grass and awash with wildflowers in the spring, coastal prairie is found on the heavy soils of coastal terraces at Pomponio State Beach, San Mateo County. Todd Keeler-Wolf

Inset: California oat grass in the coastal prairie. Deborah Hillyard

Recreation Area. Another declining species found associated with the fringes of coastal salt marsh is the hemi-parasitic salt marsh bird's beak (*Cordylanthus maritimus* ssp. *maritimus*). This species has fared better than the sea-blite, and currently ranges from San Francisco Bay Area south to Baja California, Mexico.

Coastal Terrace Prairie

Along the immediate coast, areas of marine sediments of Miocene origin that have been uplifted in place remain generally flat and somewhat terraced. The highest (and furthest inland) terraces were the first to be uplifted and are therefore the oldest. In some areas, the terraces are overlain by sand dunes of more recent origin, although dunes on the higher terraces may be quite ancient. The array of vegetation over these uplifted marine sediments is reflective of the soil texture and the time the soils have spent weathering in place.

Generally, the soils underlying coastal terrace prairie remain moist well into the dry season, as they are deep soils composed of clays and other fine textured materials, or underlain by a hard pan, where summer fog supplements the seasonal precipitation along the coast. South of the San Francisco Bay area, this prairie is common on marine terraces near the coast as well as moist sites occurring on finer-grained soils away from the immediate coast but within fog range. This habitat extends as far south as Harmony Headlands on the San Luis Obispo County coast.

Coastal terrace prairie on the Central Coast is most commonly characterized by the presence of California oat grass (*Danthonia californica*) prairie alliance, a native perennial bunchgrass fairly tolerant of light grazing and compacted soil. North of San Francisco Bay, California oat grass prairie occurs primarily away from the immediate coast, on balds and in the interior North Coast Range. On the Central Coast, reliable associates include annual and herbaceous perennial species which contribute to spring wildflower displays in many years. These include species such as California buttercup (*Ranunculus californicus*), Johnny-jump-up (*Viola pedunculata*), dwarf checkerbloom (*Sidalcea malviflora*), goldeneggs suncups (*Camissonia ovata*), and plants with bulb-like roots such as wavyleaf soap plant (*Chlorogalum pomeridianum*), Douglas iris (*Iris douglasiana*), and blue-eyed grass (*Sisyrinchium bellum*).

Other native perennial grass species, often definitive of grasslands in more inland sites, can also be important in the central coastal prairie, including purple needle grass (*Nassella pulchra*), red fescue (*Festuca rubra*), blue wild rye (*Elymus glaucus*), nodding needle grass (*Nassella cernua*), and meadow barley (*Hordeum brachyantherum*). Notably, coyote brush (*Baccharis pilularis*) is one of the most common native associates in the coastal terrace prairie. This species' presence is indicative of the dynamic tension between grassland and woody vegetation, perhaps mediated by salt air, grazing/browsing, and fire.

Native species diversity in coastal prairie of the Central Coast is nearly twice that of the nutrient-poor serpentine grasslands (discussed later in this chapter). It would appear that the high

Top: The dynamic interplay between coyote brush and coastal herbaceous vegetation at Pescadero State Beach.

Middle: Coastal prairie with tufted hair grass and Douglas iris, Marin County.
TODD KEELER-WOLF PHOTOS

Right: Uncharacteristic for manzanita, Hearst's manzanita is found sprawling in coastal prairie (here in front of the much larger—and upright—coast live oak) on San Luis Obispo County's north coast.
DEBORAH HILLYARD

Above: Coastal prairie occurs in small patches in a complex mosaic with coastal bluff scrub on slopes and terraces adjacent to the coast; here above Little Sur River, Monterey County.
JEFF BISBEE

Insets: Dwarf goldenstar (left) and Arroyo de la Cruz mariposa lily (right) are found only in a very limited area north of Arroyo de la Cruz in San Luis Obispo County.

Below: Coastal prairie is found away from the coast in areas within reach of the summer fog. Two rare species, Santa Cruz tarplant (left) and San Francisco popcorn flower (right), grow at Graham Hill in Santa Cruz County.
DEBORAH HILLYARD PHOTOS
LILY: DAVE HACKER

species diversity and high cover of the coastal prairie are effective at keeping exotic species at bay (Stromberg et al. 2001). However, a number of declining species are associated with California oat grass prairie. Their decline is indicative of a general loss of this vegetation type, due in part to the agricultural value of the soils that typically underlay coastal terrace prairie, and also because these terrace areas are desirable locations for suburban and urban development. At-risk herbaceous species include San Francisco popcorn flower (*Plagiobothrys diffusus*), Hickman's cinquefoil (*Potentilla hickmanii*), and Santa Cruz tarplant (*Holocarpha macradenia*), occurring from the San Francisco Bay to the Monterey Bay areas. Santa Cruz clover (*Trifolium buckwestiorum*), Pacific Grove clover (*Trifolium polyodon*), and coastal dunes milkvetch (*Astragalus tener* var. *titi*) are local endemic species in the vicinity of Monterey Bay. The Ohlone tiger beetle (*Cicindela ohlone*) is a very narrow endemic found in high-quality coastal terrace prairie in Santa Cruz County; the fate of this species is very closely tied with the future of the prairie it calls home.

Endemic plants that occur in coastal prairie from Monterey Bay to San Luis Obispo County include Hickman's onion (*Allium hickmanii*), dwarf goldenstar (*Bloomeria humilis*), Cambria morning-glory (*Calystegia subacaulis* ssp. *episcopalis*), Arroyo de la Cruz mariposa lily (*Calochortus clavatus* var. *recurvifolius*), and adobe sanicle (*Sanicula maritima*). In coastal terrace prairie on clay soils north of Arroyo de la Cruz in this area

are found two rare species of genera elsewhere typical of chaparral: Hearst's manzanita (*Arctostaphylos hookeri* ssp. *hearstiorum*), and maritime ceanothus (*Ceanothus maritimus*). Both are local endemic shrub species that occur in a naturally prostrate form in the prairie.

Pacific reed grass (*Calamagrostis nutkaensis*) meadows alliance and tufted hair grass (*Deschampsia cespitosa*) meadows alliance, along with a number of other prairie types, are found on the immediate coast north of San Francisco Bay, extending to British Columbia. Although these two species occur as far south as San Luis Obispo County, only the latter characterizes a prairie community; tufted hair grass meadows occur along the north San Luis Obispo coast in a very narrow band, literally within throwing distance of the bluff edge. Tufted hair grass can be a wetland species; coyote thistle (*Eryngium armatum*), also sometimes a wetland species, is found in the tufted hair grass meadows as well—an indication of very heavy soils.

Coastal Scrub

Coastal scrub is typically comprised of relatively short-lived, low shrubs, adapted to wind and salt air off the ocean by way of a variety of morphological strategies, including drought-deciduous, evergreen, and winter-deciduous leaves. This vegetation is multifarious as represented on the slopes, terraces, and dunes—and the species array depends in large part on proximity to the coast, underlying soil texture, and disturbance regime (Ford & Hayes 2007). Although termed coastal scrub, this vegetation

Above: The density of the vegetation is higher in areas protected from the wind, such as the dune scrub dominated by mock heather in the foreground; the dunes in the background, closer to the ocean, are less stable and have patches of bare sand, here along Morro Bay.
Andrea Pickart

The rare surf thistle (inset left) and beach spectaclepod (inset right) thrive in open dune patches. Guadalupe Dunes, San Luis Obispo County.
Deborah Hillyard photos

extends well inland along corridors of maritime climate influence, including coastal fog and cool moist marine air traveling on the prevailing winds. Coastal scrub in the central coast has a number of dominant or co-dominant species, such as California sagebrush (*Artemisia californica*), coyote brush, California coffee berry (*Frangula californica*), blue blossom (*Ceanothus thyrsiflorus*), and poison oak (*Toxicodendron diversilobum*). Other species such as black sage (*Salvia mellifera*), mock heather (*Ericameria ericoides*), several shrubby buckwheats (*Eriogonum fasciculatum, E. latifolium* and *E. parvifolium*), bush monkeyflower (*Diplacus aurantiacus*), golden yarrow (*Eriophyllum confertiflorum*) and seaside woolly sunflower (*Eriophyllum staechadifolium*), yellow bush lupine (*Lupinus arboreus*), and silver dune lupine (*L. chamissonis*) can also be co-dominant. This scrub is another example of the mixture of northern and southern influences of the Central Coast.

Shrubs in this community can be evergreen such as coyote brush, California coffee berry, and blue blossom, or they can be winter deciduous, such as oceanspray (*Holodiscus discolor*) or poison oak. These species typically prefer more moist or foggy exposures than the drought-deciduous shrubs such as the true sages, California sagebrush, or the buckwheats.

Top: Coastal shrubs nearest the coast include seaside woolly sunflower, buckwheats, California sagebrush and many others, Monterey County.

Middle: A mixture of coastal shrubs from coyote brush, poison oak, California sagebrush, California coffee berry, and bush monkeyflower occur along steep, drier slopes at Big Sur with coast live oak, tanoak, and redwood.
Julie M. Evens photos

Left: Inland coastal shrubs include California sagebrush and black sage, near Livermore at Corral Hollow with blue oak.
Pete Veilleux

The more drought hardy species increase relative to the previous group moving inland and south along the coast where precipitation and the cooling effects of persistent summer fog decrease.

Closer to and along the beach, in looser, wind-blown sand, herbaceous perennial species such as sand verbenas (*Abronia latifolia* and *A. umbellata*), beach sagewort (*Artemisia pycnocephala*), and beach bur-sage (*Ambrosia chamissonis*) are dominant or co-dominant in the dune mat alliance. Other herbaceous perennials, such as deer weed (*Lotus scoparius*), California croton (*Croton californicus*), and rush-rose (*Helianthemum scoparium*), are also found in more open or disturbed areas.

The various coastal scrubs overlap in terms of component species, linked through a disturbance continuum. The active dune environment is very dynamic, and perhaps because of this, low in organic matter and nutrients with a lower water holding capacity. Areas subject to more frequent disturbance, including those closest to the beach, as well as those areas subject to blow-outs, human disturbance, fire and the like, will support species which are less woody, shorter lived, and able to survive in a depauperate environment. In the absence of frequent disturbance, woody longer-lived species are able to move in and be successful, and will in turn cycle more organic material and nutrients back into the substrate. At the more stable end of this disturbance continuum, but still subject to periodic perturbation, is maritime chaparral, which includes larger woodier shrub species in addition to many of the same species characteristic of the coastal scrub communities.

Near Arroyo de la Cruz in northern San Luis Obispo County, on marine terraces nearest to the ocean, is found a windswept coastal terrace scrub dominated by seaside daisy (*Erigeron glaucus*), Pacific gum plant (*Grindelia stricta* var. *platyphylla*), and seaside woolly sunflower; with California sagebrush, sea thrift (*Armeria maritima*), and coyote brush. This coastal terrace scrub is on sandier soils than those which support coastal terrace prairie; it is here where the rare compact cobwebby thistle (*Cirsium occidentale* var. *compactum*) is found.

Coastal dune scrub is associated with a number of at-risk species. Pacific Grove in Monterey County (as well as other dune areas north of San Francisco Bay) hosts the state and federally endangered Tidestrom's lupine (*Lupinus tidestromii*), a shrubby lupine that can be co-dominant. Other declining species—mostly annual and perennial herbs—are associated with coastal dune scrub, including Menzies'

Above: Dune mat with Wight's paintbrush (Castilleja wightii) *and beach sagewort at Pescadero State Beach.*
Andrea Pickart

Below: An early migrant on the Central Coast, Allen's hummingbird (Selasphorus sasin) *breeds only along a narrow strip of coastal California and just into southwestern Oregon.*
Brad Schram

Above: On a windswept coastal terrace, scrub dominated by seaside daisy and golden yarrow is found on sandier soils than those which support coastal terrace prairie; the rare compact cobwebby thistle (inset) occurs here. North Coast, San Luis Obispo County.

Right: Brilliant white sand characterizes the dunes in Pebble Beach, which support dune scrub (foreground) as well as Monterey pine forest on the older more stabilized dunes.

Inset: Menzies' wallflower's three subspecies are found only in dunes of Humboldt, Mendocino and Monterey Counties.
Deborah Hillyard photos

wallflower (*Erysimum menziesii*), beach wallflower (*E. ammophilum*), sand gilia (*Gilia tenuiflora* ssp. *arenaria*), beach layia (*Layia carnosa*), beach spectaclepod (*Dithyrea maritima*), Nipomo lupine (*Lupinus nipomensis*), as well as surf thistle (*Cirsium rhothophilum*) and La Graciosa thistle (*Cirsium loncholepis*).

Maritime Chaparral

Chaparral communities, widespread throughout California, are characterized by woody, thick-leaved, evergreen shrubs, adapted to our Mediterranean climate. Along the coast, the influence of summer fog serves to mitigate summer heat, resulting in a longer natural fire return interval for the entire landscape. Chaparral in this area is termed "maritime" due to marine influence, and is characterized by a variety of local endemics in the genera *Arctostaphylos* and *Ceanothus*. Many are considered to be "obligate seeders," as they regenerate

Above: Maritime chaparral is found within reach of the Central Coast's characteristic summer fog, here along Point Sal, Santa Barbara County.
Jeff Bisbee

Left: Wildfires are a natural process in maritime chaparral, and promote regeneration of chaparral species. Large wildfires like the 1994 Highway 41 fire on West Cuesta Ridge, San Luis Obispo County, are increasingly uncommon due to the emphasis on controlling fires at the smallest possible acreage to minimize human casualties and property damage along the Central Coast.
Deborah Hillyard

Top: Monterey manzanita on Aromas formation at Fort Ord in Monterey County.
Jeff Bisbee

Above: Hooker's manzanita and Pajaro manzanita, endemic chaparral along ancient dune formations in Monterey County.
Julie M. Evens

after fire and other disturbance only by germinating from seed, rather than by way of crown sprouting from the base of the old plant. Because the desirable coastal location has resulted in residential development in areas vegetated by maritime chaparral, opportunities for both natural (wildland) and prescribed fires are diminishing. In the absence of fire, management of this community to maintain all of its characteristic species will be a challenge in the long-term. Work in the Prunedale Hills suggests that the absence of fire may favor obligate seeders over crown-sprouting manzanitas; though a long absence may eventually favor crown-sprouters such as oaks (*Quercus* spp.) and toyon (*Heteromeles arbutifolia*) over obligate seeders such as the endemic manzanitas (VanDyke et al. 2001).

Maritime chaparral is often found on ancient dunes comprised of sand deposited by the wind on marine terraces adjacent to the coast and subsequently uplifted. These ancient dunes have weathered to include substantial organic matter and often include clay lenses or layers. This geomorphic setting can be found throughout the Central Coast, and in each location one or more species of manzanita can be found in combination with chamise (*Adenostoma fasciculatum*), Monterey ceanothus (*Ceanothus rigidus*), sand buckbrush (*C. cuneatus* var. *fascicularis*), toyon, California coffee berry, coastal silk tassel (*Garrya elliptica*), holly leaf cherry (*Prunus ilicifolia*), and other shrub species commonly found in coastal scrub communities, such as bush monkeyflower, coyote brush, California sagebrush, black sage and deer weed.

In Monterey County's Elkhorn Slough watershed, both Hooker's manzanita (*Arctostaphylos hookeri*) chaparral alliance and Pajaro manzanita (*A. pajaroensis*) chaparral alliance are characteristic. Here maritime chaparral

is underlain by ancient consolidated dunes of the Aromas formation. On the former Fort Ord, between the cities of Marina and Monterey, Monterey manzanita (*A. montereyensis*) chaparral alliance is found primarily on the (mid-Pleistocene) Aromas formation, while sandmat manzanita (*A. pumila*) chaparral alliance occurs on relatively newer (late-Pleistocene) Baywood sands. On the Monterey and Point Lobos peninsulas, sandmat manzanita chaparral is characteristic on Baywood sands, and Hooker's manzanita chaparral is found on marine terraces. Both types can also occur as the understory to Monterey pine (*Pinus radiata*) forest.

Maritime chaparral includes a number of other endemic and characteristic shrub species in addition to the endemic manzanitas. In Monterey County, species include Monterey ceanothus and Eastwood's goldenbush (*Ericameria fasciculata*). Also frequently associated are other declining endemic herbaceous species such as Yadon's piperia (*Piperia yadonii*), sand gilia (*Gilia tenuiflora* ssp. *arenaria*), seaside bird's-beak (*Cordylanthus rigidus* var. *littoralis*), and Monterey spineflower (*Chorizanthe pungens* var. *pungens*). This area also provides habitat for a local endemic found in sandy soils—the black legless lizard (*Anniella pulchra nigra*), a sub-species of a more wide ranging species, the silvery legless lizard (*A. pulchra pulchra*). It is thought the black legless lizard's black coloration helps with thermoregulation in the cooler coastal environment of Monterey Bay.

In San Luis Obispo County, maritime chaparral is found on ancient dunes of Baywood sands in the vicinity of Los Osos, on the south shore of Morro Bay. Morro manzanita (*Arctostaphylos morroensis*) is the characteristic endemic species, co-occurring on some sites with another more wide-ranging endemic,

Above: Morro manzanita characterizes the maritime chaparral on ancient dunes south of Morro Bay estuary. While it once had a distribution of around 1,200 ha, it is now less than 400 ha, due primarily to urbanization. Morro Bay State Park, San Luis Obispo County.
Jeff Bisbee

Below: A denser stand of Morro manzanita, showing its typical habitat.
Deborah Hillyard

Above: Burton Mesa chaparral with sand mesa manzanita, La Purissima manzanita, and wedge leaf ceanothus, Santa Barbara County.
Jeff Bisbee

Below: Maritime chaparral on the Zayante formation can include silverleaf manzanita and ponderosa pine, and has naturally occurring openings on ridgetops dominated by a number of rare and disjunct herbaceous species.
Jodi McGraw

Pecho manzanita (*A. pechoensis*). The state and federally endangered Morro Bay kangaroo rat (*Dipodomys heermanni morroensis*) is associated with Baywood sands, as well as the locally endemic Morro blue butterfly (*Icaricia icarioides moroensis*). Further south on Nipomo Mesa, sand mesa manzanita (*A. rudis*) is characteristic of the maritime chaparral which has been almost completely extirpated by clearing for agriculture, golf courses and mini-ranchettes.

On Burton Mesa near Lompoc, in Santa Barbara County, maritime chaparral is underlain by soils in the Marina series, developed on aeolian sandy coastal-plain materials. The characteristic manzanitas here are La Purissima manzanita (*Arctostaphylos purissima*) and sand mesa manzanita. Of note is the disjunct occurrence on Burton Mesa of seaside bird's-beak, also found in maritime chaparral in Monterey County. In San Luis Obispo and Santa Barbara Counties, the rare sand buckbrush is frequently associated with the maritime chaparral.

In some areas—notably in the San Andreas area in Santa Cruz County, on the former Fort Ord in Monterey County, and in Los Osos, San Luis Obispo County—a charming "elfin" form of coast live oak underlain by ancient dune soils occurs in a mosaic with maritime chaparral. The elfin forest oaks are multi-

stemmed, rather than having one main trunk typical of coast live oaks elsewhere. As the elfin forest name implies, the oaks are rather short in stature and with the canopy touching the ground, they evoke a magical woodland environment.

In contrast with other occurrences of maritime chaparral located on ancient sand dunes, the Zayante Sandhills region of Santa Cruz County originated as marine deposits from the Miocene, with thick beds of sand dollars and other fossils that have become indurated or hard. The Zayante Sandhills supports Ben Lomond or silverleaf manzanita (*Arctostaphylos silvicola*) chaparral alliance, characterized by the endemic manzanita in association with other species typical of maritime chaparral elsewhere, and may include other species of manzanita such as wooly manzanita (*A. tomentosa* ssp. *crinita*) and glossy leaf manzanita (*A. sensitiva*). In addition to silverleaf manzanita three rare herbaceous endemics are found here: Ben Lomond spineflower (*Chorizanthe pungens* var. *hartwegiana*), Ben Lomond buckwheat (*Eriogonum nudum* var. *decurrens*) and Santa Cruz wallflower (*Erysimum teretifolium*). Several sensitive animal species associated with the Zayante sands include the Santa Cruz kangaroo rat (*Dipodomys venustus venustus*), in severe decline and last known from only one site near Felton; and two federally endangered endemic insects, Zayante band-winged grasshopper (*Trimerotropis infantilis*) and Mount Hermon June beetle (*Polyphylla barbata*).

The Zayante Sandhills also support a highly disjunct coast range population of ponderosa pine (*Pinus ponderosa*), and stands of Santa Cruz cypress (*Hesperocyparis abramsiana*), unusual in the sea of redwood (*Sequoia sempervirens*) and mixed evergreen forests found on

Above: Coast live oak elfin forest, Monterey County.
JEFF BISBEE

Below: Mount Hermon June beetle is a ground-dwelling scarab beetle, endemic to the Zayante Sandhills in maritime chaparral and ponderosa pine forest, here at Quail Hollow Quarry Conservation Area.
JODI MCGRAW

Above: Monterey pine forest is found in areas of ocean upwelling associated with coastal summer fog, here at Point Lobos State Natural Reserve, Monterey Peninsula.

Below: Monterey pygmy cypress in Point Lobos State Natural Reserve.
JEFF BISBEE PHOTOS

finer-textured, more nutrient-rich soils off the Zayante sands. Additionally, several species found primarily in coastal scrub are relicts here, including California sea thrift (*Armeria maritima* ssp. *californica*), mock heather, and beach sagewort, among others (Morgan 1980).

Closed-cone Pine and Cypress Stands

Although Monterey pine may be the most widely planted tree in the world, all commercial stock originates from five native coastal Monterey pine forests. Three of these are found in areas influenced by ocean upwelling and extensive summer fog on the Central Coast of California: near Swanton, Santa Cruz County; on the Monterey and Point Lobos peninsulas, Monterey County; and in and around Cambria, San Luis Obispo County. The other two native forests are on islands off the coast of Baja California, Mexico.

Native mainland Monterey pine forests are found on marine terraces of varying age and soil characteristics. Many stands are on terraces with soils that are acidic and poorly drained, underlain by a claypan or hardpan and associated with wetland species. Monterey pine forest can also be found on well-drained soils, as well as older dunes; these dunes have an accumulation of organic

material and, occasionally, thin bands of clay which have been transported over time from the upper layers of dune (Cylinder 1995).

These pine forests often include coast live oak and, on the Monterey Peninsula, may be associated with Monterey pygmy cypress (*Hesperocyparis goveniana*) or bishop pine (*Pinus muricata*). Monterey pine forest may have an understory of chaparral shrub species, coastal prairie species, wetland species or no understory at all, just a carpet of pine needles. Bishop pine forest also occurs sporadically in patches on the Central Coast, with similar associates, and like Monterey pine, requires summer fog.

On the Monterey and Point Lobos peninsulas are marine terraces that have been uplifted and weathered to produce shallow, extremely acidic soils underlain by a clay pan. Here Monterey pygmy cypress stands persist on sites that pond in winter and bake in the high summer heat. These extreme conditions result in individuals that are dwarfed on the most severe sites, growing to only 5 meters (m) in height as compared to others that may reach up to 50 m on more fertile sites.

Along the immediate coast in Point Lobos State Natural Reserve and Crocker Grove in Pebble Beach, areas with well-drained soils formed from decomposed granite bedrock support Monterey cypress (*Hesperocyparis macrocarpa*) stands alliance (Cylinder 1995). While this species has a tendency to become naturalized and has been planted widely along the coast, these two areas are the only locations where native stands occur.

All closed-cone pines and cypresses have serotinous or semi-serotinous cones. The seeds, stored in the cones that are found in the tree canopy, are released typically under conditions of heat sufficient

Above: Though widely planted and extensively naturalized, native stands of Monterey cypress occur in only two locations in the world, both in Monterey County; this stand is in Crocker Grove.
Jeff Bisbee

Below: Southern sea otter (Enhydra lutris nereis) *is a keystone species of the nearshore kelp forest ecosystem, Monterey Bay.*
Brian Barnes

Seven years after the 1987 fire in Pebble Beach, which burned 160 acres, young bishop and Monterey pine emerged from maritime chaparral, which was first to develop post-burn. Monterey Peninsula, Monterey County.
Deborah Hillyard

to break down the resin that binds the cones shut. Cones may also open with age, although the seeds of such aged cones often have low viability. Air temperatures sufficiently hot enough to open the cones are unusual along the coast, but do occur occasionally. Fire is especially effective at opening the cones, whose seeds are then released into a favorable seedbed. It is, however, a delicate balance: a fire must be sufficiently hot to open the cones but not so fierce that the cones and seeds inside them incinerate. These cones also have an ingenious adaptation to fire: the resin break-down is delayed for up to several days, so that seeds do not get discharged directly into active flames. However, their viability in a soil seed bank is short lived, typically less than a year once on the ground. This presents a conservation and management conundrum unusual in plants, necessitating extra care to preclude the loss of the seed stock through our management activities, especially with prescribed fire. Potential progeny of the closed-cone conifers are not stored in the soil seed bank (typical for plants), but are typically found in the canopy of the living adult (as is more normal for animals), and seeds could be killed through an out-of-season, too-hot, or too-cool burn. These stands must be managed thoughtfully to preserve the seed bank and preclude the loss of subsequent generations.

Serpentine Communities

Snaking throughout the Franciscan formation, bands of an unusual substrate support a unique flora and vegetation. This substrate consists of thin soils, derived from serpentinite and other ultramafic rock types. The ultramafic rock was extruded at the oceanic crust typically near mid-ocean ridges and is sometimes metamorphosed. After a hundred million years or so of travelling across the ocean floor, it was scraped up onto the continental crust along the subduction zone.

Ecologists typically lump soils derived from all ultramafic substrate (not just serpentine) as serpentinite soils, and similarly, collectively refer to the plant communities that these soils support as "serpentine" vegetation. These so-called serpentinite soils are low in essential nutrients such as nitrogen, phosphorus and potassium. They have a low calcium-to-magnesium ratio and low silica content; however, they are high in toxic heavy metals (particularly nickel, but also including iron, cobalt and chromium). All of these characteristics render serpentinite soils an extremely hostile

plant-growth medium.

Most plants cannot grow in this impoverished and potentially toxic edaphic environment, though plant species' tolerance of these extreme conditions spans a broad spectrum. The stress to plants from limited rooting area due to its shallow soils, combined with the chemical challenges of the serpentinite soils, allows for the establishment and persistence of only select taxa tolerant to serpentinite soils' limitations and excesses. Plants not so adapted are effectively excluded. Endemic plant species more or less restricted to these types of soils are considered to be specialty taxa (i.e. species, sub-species and races/ecotypes), which have evolved to be able to persist and do well by way of specialized adaptations. Many have biochemical or physiological mechanisms that allow them to sequester particular elements to exclude or facilitate their uptake. Mycorrhizal associations may also be important in development of such tolerance (Hopkins 1987). It also appears that at some sites, drought tolerance is equally important, as reflected in a suite of morphological adaptations such as xeromorphic foliage and/or small stature (Brady et al. 2005). Taxa endemic to serpentinite substrates are often unique. While making up only a small portion of the vegetative cover, they are characteristic and a relatively constant component. Interestingly, a major threat to serpentine vegetation is the deposition of excess nitrogen originating primarily from vehicle emissions in urban areas, which renders the soil much more ordinary and available to colonization by non-serpentine species.

Common or characteristic species that may be found on both serpentinite and non-serpentinite substrates include foothill pine (*Pinus sabiniana*), Coulter pine (*P. coulteri*), knobcone pine (*P. attenuata*), birch leaf mountain mahogany (*Cercocarpus montanus* var. *glaber*),

In addition to direct loss due to urbanization, serpentine grasslands like this one overlooking the Santa Clara Valley are vulnerable to a more insidious threat, encroachment by non-native grasses due to the deposition of excess nitrogen originating in vehicle emissions. Coyote Ridge, Santa Clara County.
Julie M. Evens

Top: Areas on ridges and rock outcrops where the soil is so thin as to be virtually non-existent are termed serpentine barrens. Sierra Azul Preserve, Santa Clara County.
Julie M. Evens

Above: Sargent cypress stands can be found on serpentine ridgetops, here with bishop manzanita in the foreground; periodic fire will facilitate germination of the seeds of both species. West Cuesta Ridge, San Luis Obispo County.
Deborah Hillyard

chamise, wedge leaf ceanothus (*Ceanothus cuneatus*), silk tassel (*Garrya* spp.), capitate gilia (*Gilia capitata*), chia (*Salvia columbariae*), and bracken fern (*Pteridium aquilinum*). However, some may be ecotypically differentiated into serpentine-tolerant and -intolerant races. Common and characteristic species of non-serpentinite soils of the Central Coast, which do not appear to "cross-over" to serpentinite soils, include ponderosa pine, coast live oak, California buckeye (*Aesculus californica*), and Eastwood's manzanita (*Arctostaphylos glandulosa*).

The distribution of vegetation on serpentinite soils is reflective of the same parameters that influence distribution of vegetation on non-serpentinite sites, such as soil depth, soil texture, soil chemistry, aspect, and slope. On the Central Coast, serpentine herbaceous communities are commonly found on steep slopes and ridges with abundant protruding rocks, as well as on shallow slopes and alluvial deposits (the latter two geomorphic locations are preferred north of the San Francisco Bay area). Herbaceous communities tend to range from denser grasslands on deeper soils, with increasing dominance by herbaceous annuals and perennial wildflowers on the thinner soils, with a specialized community found on rock outcrops and ridges, called "serpentine barrens" by Kruckeberg (1984). Serpentine chaparral is found on sites with steep slopes and shallower soils, often with a south or west aspect that renders the site even more arid. Sargent cypress (*Hesperocyparis sargentii*), another closed-cone conifer type, is routinely found on serpentinite ridgetops (including Cuesta Ridge in San Luis Obispo County), which is somewhat unusual given the shallow soils and general aridity of such sites. Other woodlands characterized by broadleaf evergreen and conifer species are found on more mesic serpentine sites: on north facing slopes, areas of deeper soils, or anywhere there is more soil water holding capacity. Serpentine seeps, tucked into the landscape here and there, occasionally bring a little moisture to the surface.

In serpentine herbaceous communities, native perennial grass species can range in density from present-but-not-abundant to almost pure stands. Characteristic grassland types include purple needle grass grasslands in coastal areas, and foothill needle grass (*Nassella lepida*) and curly blue grass (*Poa secunda*) grasslands in more inland areas. Other native perennial grass species can also be present, even abundant, depending on whether the site is grazed or not—include Torrey's melic grass (*Melica torreyana*), big squirreltail (*Elymus multisetus*), and prairie junegrass (*Koeleria macrantha*). Small fescue (*Vulpia microstachys*), a native annual grass, is also common and characteristic of these communities.

It should be noted that serpentine grasslands (as well as herblands and barrens) are best known not for their grass species, but rather for their annual and perennial forbs that provide iconic wildflower displays. Like the grasses, many of these showier species are common off serpentinite—species like goldfields (*Lasthenia californica* and *L. gracilis*), lupines (*Lupinus* spp.), and yarrow (*Achillea millefolium*). It is thought that these local representatives of more ubiquitous species have adaptations that result in some level of tolerance for serpentinite conditions. Also a number of herbaceous serpentine endemics are locally rare and of particular interest to plant enthusiasts. San Mateo thornmint (*Acanthomintha duttonii*) occurs in only a few locations in heavy clay soils in San Mateo County. Presidio clarkia (*Clarkia franciscana*) is found on the Presidio of San Francisco and at Redwood Regional Park in the East Bay. Metcalf Canyon jewelflower (*Streptanthus albidus* ssp. *albidus*) and Santa Clara Valley dudleya (*Dudleya abramsii* ssp. *setchellii*) are endemic to Santa Clara County. Most beautiful jewelflower (*S. albidus* ssp. *peramoenus*) ranges throughout the East Bay south to San Luis Obispo. San Luis mariposa lily (*Calochortus obispoensis*) is endemic to San Luis Obispo County and one of several mariposa lilies restricted to serpentinite substrates.

Serpentine chaparral is also found throughout the Central Coast area. Although this chaparral includes some common species such as chamise, toyon, and bigberry manzanita (*Arctostaphylos glauca*), there are also numerous species with an affinity for serpentinite substrates. These include leather oak (*Quercus durata*) and Congdon's silk-tassel (*Garrya congdonii*), as well as local and rare serpentine specialists such as Franciscan manzanita (*Arctostaphylos franciscana*), Presidio manzanita (*A. montana* ssp. *ravenii*), coyote ceanothus (*Ceanothus ferrisiae*), bishop manzanita (*A. obispoensis*), Cuesta Pass checkerbloom (*Sidalcea hickmanii* ssp. *anomala*), San Luis Obispo sedge (*Carex obispoensis*), and Santa Ynez ceanothus (*C. papillosus* var. *roweanus*).

Serpentine seeps provide the perfect microhabitat for several related endemic native thistles. Fountain thistle (*Cirsium fontinale* var. *fontinale*) is known from the eastern foothills of the Santa Cruz Mountains in San Mateo County. Mount Hamilton thistle (*C. fontinale* var. *campylon*) is found on serpentinite throughout southeastern Santa Clara County. Chorro Creek bog thistle (*C. fontinale* var. *obispoense*) is found in a number of locations from Cambria to the Irish Hills in San Luis Obispo County.

Despite the San Francisco Bay area's intensive urbanization, excellent examples of conserved landscapes, which also preserve serpentine vegetation, include

Above: Serpentine chaparral is characterized by a number of species which are common in non-serpentine chaparral communities (but probably in local races which are adapted to be serpentine specialists) as well as species which are serpentine endemics. Mount Diablo State Park, Contra Costa County.
DEBORAH HILLYARD

Below: A characteristic resident of chaparral, the California thrasher (Toxostoma redivivum) *is often heard singing and can occasionally be seen, perched high on a shrub or scurrying on the ground; they are found only in California and northern Baja California.*
BRAD SCHRAM

Above: Seeps provide year-round moisture in otherwise dry summer hillsides, supporting endemic serpentine seep thistles. Edgewood, San Mateo County.
Deborah Hillyard

Below: Mount Hamilton thistle in bloom, Coyote Ridge, Santa Clara County.
Julie M. Evens

Golden Gate National Recreation Area and a small area in the Presidio of San Francisco. San Mateo County also offers Edgewood County Park, Pulgas Ridge/ Crystal Springs Reservoir, and Jasper Ridge Biological Preserve. Alameda County's conservation areas include Redwood Regional Park and Cedar Mountain Ridge; Contra Costa County has Mount Diablo State Park. Santa Clara County's preserved areas include open space on Coyote Ridge, Santa Teresa County Park, Almaden Quicksilver County Park, and Calero Reservoir. Other prime areas include San Benito Mountain and New Idria in San Benito County; and in Monterey County, Chew's Ridge in the Los Padres National Forest, southern Big Sur, and Fort Hunter Liggett.

San Luis Obispo County includes Hearst Ranch (private) and Cuesta Ridge Botanical Special Interest Area in the Los Padres National Forest. A complex of protected areas in the City of San Luis Obispo, including Laguna Lake Open Space, Reservoir Canyon Natural Reserve, Johnson Ranch Open Space, Irish Hills Natural Reserve, Islay Hill Open Space and South Hills Open Space, demonstrate successful conservation efforts. Vandenberg Air Force Base, Figueroa Mountain in the Los Padres National Forest, and the Sedgwick Natural Reserve (managed by the University of California Santa Barbara Natural Reserve System) in northern Santa Barbara County represent significant habitat preservation efforts.

Conservation

Conservation of native vegetation has occurred throughout the Central Coast and Coast Ranges, including

San Bruno Mountain State and County Park at the northern end of the Santa Cruz Mountains is an incredible island of open space in the midst of the urbanized San Francisco Peninsula. It preserves habitat for several rare plants including San Bruno Mountain manzanita (Arctostaphylos imbricata)*, Montara manzanita* (A. montaraensis)*, and Pacific manzanita* (A. pacifica)*.*
JEFF BISBEE

the establishment of open space, parks, preserves, and conservation easements. Effective long-term conservation should continue to include active management strategies, such as well-timed grazing and/or prescribed fire, to curb non-native plant invasion.

A number of stressors threaten the long term viability of the ecosystems of the Central Coast, and indeed all of California. They are all, in one way or another, human related. The most immediate threats come from urbanization, which results in not only conversion of native vegetation to human landscapes, but also contributes to fragmentation and the loss of natural corridors, modification to the ozone in the air above our cities, deposition of nitrogen (a product of internal combustion engines), alteration of the natural fire return interval, introduction of aggressive non-native species, depletion of the water table, and the loss of natural pollinators and herbivores.

Conservation at the landscape level will be most effective at maintaining our component communities over the long term; that would include not only conservation of lands on the ground, and the plants and animals which comprise the communities, but also maintenance of the natural processes that help to shape and support the communities within the landscape. The future of any conserved area also relies on fundamental changes by individuals, making thoughtful choices which will cumulatively help to increase the viability of our planet.

DEBORAH HILLYARD

CHAPTER ELEVEN

Southern California Coast

The coast of southern California epitomizes what much of the world considers to be "California." Warm sunny valleys, extensive beaches, and rugged coastlines typify the topography, all of which is enhanced by the backdrop of high snowy mountains. This California "scene" occupies less than 10% of the entire state. Before grazing, agriculture, and development altered the landscape early Spanish explorers commented on fields of wildflowers and verdant vegetation (Minnich 2008). Enticed by mild climate and an elaborate system of aqueducts, people colonized the region in droves in the early twentieth century. The south coastal portion of California is now the most populous part of the state—currently about 17 million people live in this area.

Despite its large population, California's southern coast is still known for its wealth of biodiversity underscored by the range of vegetation occurring here. Fortunately a significant portion of native vegetation is preserved in a series of National Forests, National Parks, State Parks, and county parks. San Nicolas and San Clemente Islands are still managed by the U.S. Navy, and several military bases such as Vandenburg, Port Hueneme, Seal Beach, and Camp Pendleton on the mainland represent de facto preserves. Beautiful coastlines abound, from the wild unfettered wilderness of eight Channel Islands to the

Above: The rugged coastline of Santa Cruz Island preserves many endemic plant species.
TODD KEELER-WOLF

Opposite: The hills above South Laguna, Laguna Beach, exhibit a diversity of coastal sage scrub and maritime chaparral, along with encroaching development.
ALLAN SCHOENHERR

Torrey Pines State Natural Reserve along the mainland coast preserves unique features including the rarest pine tree in our nation, Torrey pine, amongst coastal scrub and maritime chaparral.
Jeff Bisbee

"American Riviera" of Santa Barbara to iconic Laguna Beach in Orange County. The rugged hearts of the Santa Monica, San Gabriel, and Santa Ana Mountains still reflect the region's wild nature.

The climate associated with coastal southern California is typically maritime Mediterranean. Total annual precipitation is semi-arid, seldom exceeding 25 cm per year, but persistent springtime and morning summertime fog tends to keep evaporation rates low. Proximity of the ocean tends to moderate temperature extremes. Distribution of plant communities to inland or upland sites tends to be limited by cold temperatures. The region is typically frost-free.

Befitting an exemplary Californian microcosm, the vegetation of the south coast contains some of the best developed and most diverse of California's endemic Mediterranean vegetation. Coastal Sage Scrub, Chaparral, and warm-temperate woodlands grow best in the mild climate of the southern California coastal zone. In addition to several widespread woodland alliances the South Coast also has unusual rarities such as the Engelmann oak (*Quercus engelmannii*) woodland, island oak (*Q. tomentella*), Catalina ironwood (*Lyonothamnus floribundus*), and California walnut (*Juglans californica*) groves, and Torrey Pine (*Pinus torreyana*) stands. More kinds of coastal sage scrub (23 different alliances) and chaparral (27 different alliances) occur within this ecoregion than any other in California.

In good El Niño years annual and perennial grasses and wildflowers form showy displays in rocky and heavy soil openings, as well as in post-fire shrubland openings. Even the coastal salt marshes, which in most of California are low diversity and homogeneous vegetation types, are exceptionally diverse.

Coastal sage scrub is quite diverse, apparent here in the different colors representing different species during the dry season.
ALLAN SCHOENHERR

COASTAL SAGE SCRUB

Coastal sage scrub, a community well adapted to southern California's Mediterranean climate, probably reached its historical distribution maximum after the last glacial retreat (*ca.* 12,000 YBP) (Axelrod 1978). Since that time, anthropogenic activities such as grazing, fire, agriculture, and urban development have seriously reduced its abundance. Estimates of 85% elimination from its former extent are within reason (Westman 1981, 1982). The largest contiguous remainders of sage scrub in southern California occur in the Santa Monica Mountains, the San Joaquin Hills of Orange County, and various military lands such as Vandenberg Air Force Base and Camp Pendleton.

California coastal sage scrub is a distinct group of vegetation alliances (Sawyer et al. 2009). However, coastal sage scrub has been subdivided in several ways over the years. A well-known approach (Westman 1983) divides it into three overlapping floristic groupings—Venturan, Diegan, and Riversidian. Riversidian coastal scrub is an inland form associated with higher summer temperatures and evaporation rates. The difference between the two more coastal forms—Venturan and Diegan coastal scrub—is a function of increased southerly dryness manifested in a greater proportion of succulent species toward the south. Unlike the large evergreen deep-rooted chaparral shrubs, coastal scrub shrubs are smaller in stature with shallow spreading roots. Their summer drought strategy is to lose leaves and become dormant rather than to resist drying through thickened leaves and tapping deeper moisture sources. Many of these coastal scrub species are "facultatively deciduous"—they will quickly put out new leaves whenever water does

become available whether through winter rainfall or from added irrigation.

Many of the most xerophyllic (dry-adapted) shrub communities in southern California coastal scrub, such as brittle bush (*Encelia farinosa*) scrub and California buckwheat (*Eriogonum fasciculatum*) scrub are also found in the Mojave and Sonoran deserts. Where they occur closer to the coast, they occupy the

Top: California sagebrush and California buckwheat are two common drought-deciduous shrubs co-occurring on south-facing slopes.
TODD KEELER-WOLF

Left: Stands of the coastal brittle bush scrub occur on warm generally steep south-facing slopes usually within a few miles of the coast, here at Point Loma.
THOMAS OBERBAUER

hottest sunniest exposures—often on south-facing slopes (Schoenherr 1992, 2011). Coastal sage scrub is one the most diverse groupings of vegetation communities associated with southern California. Even during the dry season the slopes are a mosaic of different colors, each color associated with a different species. Some species grow smaller leaves during the dry season, and remain photosynthetically active. On south-facing slopes vegetation is commonly dominated by drought-deciduous shrubs including California sagebrush (*Artemisia californica*). Sages in the mint family (Lamiaceae) such as white sage (*Salvia apiana*), black sage (*S. mellifera*), and purple sage (*S. leucophylla*), are joined by coastal brittle bush (*Encelia californica*), and bush monkeyflower (*Diplacus aurantiacus*) scrub alliances. On disturbed sites, California buckwheat tends to dominate. In moister coastal areas and toward the north, the large, evergreen, coyote brush (*Baccharis pilularis*) tends to become dominant. Each of these individual species are indicators of specific conditions and are not all found together in a single stand. They sort ecologically based on their relative tolerance of soil moisture, texture, slope exposure, and frequency and type of fire and other disturbance.

At exposed well-drained localities the drought-deciduous shrubs are joined by various succulent species, including a number of cactus species. Cacti absorb water quickly when it is available and it

Top: Coast cactus scrub dominated by coastal prickly pear. TODD KEELER-WOLF

Above: Bush monkeyflower; flower color in different populations varies from white, to yellow, to orange, to red. ALLAN SCHOENHERR

Top: Chaparral yucca, California buckwheat, and scale broom dominate this alluvial scrub stand in sandy Cajon Wash. Allan Schoenherr

Above: Coastal cholla near the mouth of the Tijuana River. Todd Keeler-Wolf

is stored in its succulent stems for use during the dry season. Coastal prickly pear (*Opuntia littoralis*) scrub is the most common cactus alliance. It is particularly common in areas that have been heavily grazed or on steep rocky south-facing slopes. The bright red fruit, called *tunas* by the Spanish, is an important food for many animals including humans. Seed dispersal is facilitated through the digestive tracts of coyotes (*Canis latrans*) and foxes (*Urocyon* spp.). Sometimes coexisting with coastal prickly pear, but other times located in single-species swarms, coastal cholla (*Cylindropuntia prolifera*) is an important habitat for the coastal cactus wren (*Campylorhynchus brunneicapillus*) unique to coastal sage scrub. Local land managers refer to these alliances as coastal cactus scrub.

Chaparral yucca (*Hesperoyucca whipplei*), is a leaf succulent, with some populations reproducing by cloning, producing "pups" at the base. Other populations only reproduce by seeds (Ingram 2008). Chaparral yucca occurs in chaparral as well as coastal sage scrub, and is commonly found in habitats at the sandy mouths of rivers, such as the San Gabriel River, Big Tujunga or Cajon Wash. This version of coastal sage scrub, known as alluvial scrub, is characterized by scale broom (*Lepidospartum squamatum*)—a shrub adapted to episodic flooding.

Another group of leaf succulents found in coastal sage scrub are the live-forevers (*Dudleya* spp.). Most of them have very localized distribution in specialized habitats, particularly cliff faces and rock outcrops. There are at least 17 species in the southern California coastal area, seven species in the Santa Monica Mountains alone, and another seven in San Diego County. They are not always easy to identify as many of the species hybridize. The Channel Islands have at least 12 endemic taxa of live-forevers. The densest stands of live-forever in California also occur on the Channel Islands where they form colonies on fog-shrouded bluffs or on steep and dry rocky cliffs above the ocean, occurring with Santa Cruz Island buckwheat (*Eriogonum arborescens*) and other shrubs.

The interior or Riversidian coastal sage scrub (Rundel 2007) is ecotonal (transitional) between the coast and the cactus scrub of the Colorado Desert. Desert species such as brittle bush and tall, straight valley cholla (*Cylindropuntia californica* var. *parkeri*) are common.

Certain coastal species become replaced in interior sage scrub. For example, lemonade berry (*Rhus integrifolia*) scrub is replaced by sugarbush (*Rhus ovata*). Other species, such as white sage, become more common.

In coastal San Diego County and northern Baja California different shrubs and more succulent species appear. Many of these species are more widespread south of the border and exemplify the transition zone between the coastal true deserts of Baja—part of the Sonoran Desert—(Shreve 1951, Brown 1982) and the Californian coastal scrub. This Diegan sage scrub occurs in isolated pockets with a high cover of succulent species. These interesting and diverse "maritime succulent scrub" patches can be seen at isolated localities in the San Joaquin Hills, Cabrillo National Monument, Torrey Pines State Natural Reserve, and on the southern Channel Islands of

Top: A high density of sea lettuce (Dudleya caespitosa) *on Anacapa Island.*

Above: Valley cholla is present in coastal scrub communities that are ecotonal between the coast and cactus scrub of the deserts.

Todd Keeler-Wolf photos

Top: Coastal agave at Point Loma, just north of the Mexican border. Todd Keeler-Wolf

Above: Coastal prickly pear with its succulent leaves and bright flower. Allan Schoenherr

Santa Catalina, San Nicolas, and San Clemente. The usual cactus species are joined by golden cereus (*Bergerocactus emoryi*)—which forms clusters of upright stems with long golden spines—San Diego barrel cactus (*Ferocactus viridescens*), and California fish-hook cactus (*Mammillaria dioica*). Coastal agave (*Agave shawii*), a large leaf succulent species, has spectacular flower stalks with large clusters of yellow flowers so prolific at producing nectar that it is possible to drink the sweet nectar from plucked blossoms. Following habitat destruction, the coastal agave species only occurred naturally in California at Point Loma and Border Field State Park, just north of the Mexican border, but it has been reintroduced to Cabrillo National Monument and Torrey Pines State Natural Reserve.

The Diegan sage scrub also features about seven species of live-forevers. Several unique drought-deciduous species appear here—including a California brittle bush look-alike, San Diego goldeneye (*Viguiera laciniata*) and bushrue (*Cneoridium dumosum*), which can cause dermatitis similar to that caused by poison oak. Black sage becomes replaced by Munz's sage (*Salvia munzii*). Shrub species with succulent properties include cliff spurge (*Euphorbia misera*), and jojoba (*Simmondsia chinensis*), a desert species with fruit that is a source of a high quality wax. The succulent giant coreopsis (*Coreopsis gigantea*), common in this habitat on the Channel Islands, is also found in isolated pockets on the mainland.

Fires are an important natural factor influencing the nature of coastal sage scrub, and most of the species have evolved to survive fire. A typical interval between fires is 15 to 25 years, although the fire-return period can be shorter, particularly if coastal sage scrub has been invaded by non-native Mediterranean grasses or forbs such as black mustard (*Brassica nigra*). The first year after a fire, short-lived annual wildflowers such as phacelias (*Phacelia* spp.), lupines (*Lupinus* spp.), and California poppies (*Eschscholzia californica*) dominate the landscape (O'Leary 1989). Some species such as fire poppy (*Papaver californicum*) or whispering bells (*Emmenanthe penduliflora*) seem to germinate only after fires.

Geophytes, which return from underground storage organs, reappear in great numbers after a fire, including mariposa lilies (*Calochortus* spp.), bluedicks (*Dichelostemma capitatum*), shooting stars (*Dodecatheon clevelandii*), and blue-eyed grass (*Sisyrinchium bellum*). A large spreading grass, giant wild rye (*Leymus condensatus*), is also encouraged by fire and may dominate sites with short fire intervals. For a few years after a fire, the landscape may be dominated by short-lived perennial shrubs whose seeds have been fire-activated. Fire-following shrubs include deer weed (*Lotus scoparius*), golden ear-drops (*Dicentra chrysantha*), California brittle bush, and bush monkeyflower. In some areas larger shrubs such as bush poppy (*Dendromecon rigida*) and bush mallow (*Malacothamnus fasciculatus*) become very common. After five years or so, little visible evidence of a fire remains except for some of the skeletonized stems that project above the regenerated canopy.

Top: San Diego goldeneye and red brome (Bromus rubens) *in post-burn coastal scrub.*

Center: Invasive black mustard on old fallow fields in western San Diego County.
TODD KEELER-WOLF PHOTOS

Right: Wildflowers after a fire. Many of these species require fire to germinate.
ALLAN SCHOENHERR

Top: Laurel sumac scrub in the Santa Monica Mountains National Recreation Area.
Todd Keeler-Wolf

Above: Plants on north-facing slopes are larger and retain leaves in the dry season. Laurel sumac is frost sensitive, and its leaves show the effect of the cold temperatures.

Right: Deer weed after a fire in San Mateo Canyon, Santa Ana Mountains.
Allan Schoenherr photos

Chaparral

Chaparral is the other main scrubland formation most particularly adapted to California's Mediterranean climate. Chaparral plants in southern California typically occur at higher elevation where they receive more precipitation and are subject to colder temperatures than coastal sage scrub. An alternate strategy for shrubs facing long summer drought is to develop stiff waxy-coated evergreen leaves with deep, moisture-seeking root systems. To do this successfully in a water-limited environment, these shrubs have to grow over longer periods and develop regenerative and reproductive strategies that address perturbations to the necessary long life span of such species. However, it is interesting to note that in many situations in southern California the more successful strategy for extreme summer drought tolerance is that undertaken by the drought-deciduous, short-lived, short-rooted coastal sage scrub species.

On the other hand, large evergreen-leaved, resprouting shrubs often characterize the north-facing or protected slopes immediately adjacent to coastal scrub stands. On these "mesic" sites with more available moisture or lower evaporation rates, large evergreen shrubs such as laurel sumac (*Malosma laurina*), toyon (*Heteromeles arbutifolia*) and other evergreen shrubs such as California scrub oak (*Quercus berberidifolia*)—and its rare relative, coastal scrub oak (*Quercus dumosa*)—prevail. All of these species tend to eschew the hottest driest slopes where coastal sage scrub prevails. Frost-sensitive laurel sumac has proven helpful to ranchers seeking potential sites for citrus groves. It has been claimed that Hollywood gets its name from the presence of toyon, also known as California holly. The red berries are an important food for birds, coyotes, foxes, and in the past to Native Americans and early settlers—the Spanish would bake them with sugar to make toyon pie. Many settlers in the Los Angeles area decorated their homes at Christmastime with branches of California holly. The practice was so widespread that a 1920 law made it illegal to harvest toyon boughs. Lemonade berry, another large frost-sensitive shrub, may occur in this habitat. These shrubs are particularly common on coastal bluffs where, tolerant of salt spray, they form thickets associated with smaller drought-deciduous shrubs. In the spring, sticky red berries with a sour flavor are produced. These berries soaked in water can make a lemonade-flavored drink.

Coastal Strand and Dune Vegetation

The coastal strand and dune systems of mainland southern California are only a remnant of their pre-European extent and seriously disturbed, although they occur in a natural state on several of the Channel Islands. Only relict areas of dune mat and dune scrub remain south of Point Conception (Pickart and Barbour 2007). Notable areas where native vegetation occurs include isolated beaches from Ventura to Point Dume. El Segundo, in the Los Angeles area, has a dune system that has undergone restoration, but it contains only a fragment of its historical distribution (Mattoni 1990). A depauperate representation of native dune mat flora is also present along the strand between Newport Beach and Balboa. In San Diego County, the best examples of dune and beach vegetation occur at the mouth of the Santa Margarita River at Camp Pendleton, and vestiges of the beach communities also occur along San Diego Bay and the Tijuana Estuary. Prostrate, sprawling plants such as beach bur-sage (*Ambrosia chamissonis*), sand verbenas (*Abronia umbellata, A. maritima*), beach evening primrose (*Camissonia chei-*

Above: Lemonade berry occurs where more moisture is available and is common on coastal bluffs where it seems to tolerate the salt spray.
Todd Keeler-Wolf

Inset: Close-up detail of lemonade berry in flower and fruit.
Allan Schoenherr

ranthifolia), and beach morning-glory (*Calystegia soldanella*) occur in the foredune areas where wind disturbance is a formidable environmental issue. Silver bush lupine (*Lupinus albifrons*) is an important stabilizing plant, particularly on the Channel Islands. In disturbed areas, non-native plants such as sea fig (*Carpobrotus chilensis*) and sea rocket (*Cakile maritima*) dominate. Dune scrub, composed of low-growing shrubs such as Menzies' golden bush (*Isocoma menziesii*), mock heather (*Ericameria ericoides*), and coyote brush occur in the stabilized regions to the interior. Many of these are also components of coastal sage scrub.

Estuaries

Estuaries are wetlands where freshwater mingles with seawater. Salt marshes occur adjacent to estuaries. In combination the two are among the most threatened southern coast habitats. Statewide, 90 to 95% of our coastal wetlands have been lost to human activity such as landfills and development (Dahl 1990). The greatest impact to coastal wetlands in California has occurred in Humboldt Bay, San Francisco Bay, and in southern California (Grewell et al. 2007). In an attempt to protect the remaining habitat, numerous species have been listed as threatened or endangered in southern California. These include the endemic salt marsh bird's beak (*Cordylanthus maritimus* ssp. *maritimus*); a fish, the tidewater goby (*Eucyclogobius newberryi*); and at least three birds: the light-footed Ridgeway's rail (*Rallus obsoletus* ssp. *levipes*), Belding's savannah sparrow

*Top: Red-flowered buckwheat (*Eriogonum grande *var.* rubescens*) growing with silver bush lupine and Menzies' golden bush on stable dunes.*

Left: Beach bur-sage and sand verbena on the foredunes of the beaches of Santa Cruz Island.
Todd Keeler-Wolf photos

(*Passerculus sandwichensis*), and the least tern (*Sternula antillarum*).

Management priorities for protection and restoration have required a great deal of attention from public agencies. For example, in an attempt to classify the different kinds of estuaries, eight different types have been identified on the basis of access to tidal flushing alone (Jacobs et al. 2010). A fully functioning salt marsh requires tidal flushing, which is restricted in many southern California estuaries. The large estuaries with significant tidal flow include Tijuana Estuary, Newport Bay, Anaheim Bay, Mugu Lagoon, Carpinteria Marsh, and Goleta Slough. Estuaries with varying degrees of tidal flow such as San Mateo, Santa Margarita, San Luis Rey, Batiquitos, San Marcos, San Diego, and San Dieguito Creeks occur in San Diego County. Bolsa Chica, currently the largest estuary under restoration, had a new opening to the sea restored in August 2006, ushering in a new era of tidal flushing.

Marsh habitat doesn't end where the dry land begins. What is not apparent to the naked eye is the natural exchange of nutrients and other materials between a marsh and its surroundings. Tidal flushing stirs in oxygen and nutrients from the sea, reducing stagnation and anaerobic conditions in the sediment. Runoff from adjacent uplands carries debris and beneficial organic matter to the wetland, which eventually decomposes and becomes part of the nutrient-rich mud. Furthermore, birds that feed in the marsh may nest on land and vice versa, carrying materials to and fro in their digestive tracts and droppings.

Salt Marshes

A salt marsh, in spite of its seemingly monotonous appearance, rivals a

Top: Salt marsh, dominated by California cordgrass, in Upper Newport Bay Ecological Reserve. Allan Schoenherr

Above: Light-footed Ridgeway's rail is an endangered species found exclusively in coastal salt marshes. Jim Frazee

Top: Alkali heath is characteristic of the high marsh, here at Los Peñasquitos Lagoon.
Todd Keeler-Wolf

Above: Pickleweed marsh in Bolsa Chica Ecological Reserve, Orange County.

Inset: Long-billed curlew (Numenius americanus) *in pickleweed, Bolsa Chica Ecological Reserve.*
Allan Schoenherr photos

tropical forest in primary production. It has a high rate of photosynthesis, promoted in southern California by the keystone species California cordgrass (*Spartina foliosa*). California cordgrass marsh requires tidal flushing, so those estuarine habitats that have reduced access to the sea lack this alliance. California cordgrass roots absorb salt and water, but the salt is excreted upon its leaves making the foliage inedible to most animals.

Based on soil salinity and aeration, other salt-tolerant plants occur in rings around the salt marsh, and can be categorized into high and low salt marsh. The low marsh sustains regular tidal flooding every day. The salt water from intermittent higher tides tends to evaporate and increase the salinity in the high marsh so that a significant salt crust builds up in small depressions known as salt pannes where vegetation is largely inhibited. The relatively hot and dry climate compared to the northern California salt marshes tends to make for a sharper salinity gradient and enables a variety of salt-tolerant species not present in cooler north coastal marshes to occur.

Inland from the California cordgrass marsh is a belt of succulent plants, including pickleweed (*Sarcocornia pacifica* [*Salicornia depressa*]) mats. The name pickleweed comes from a European tradition of eating young sprigs as if they were a pickled vegetable. Saltwort (*Batis maritima*) and marsh jaumea (*Jaumea carnosa*) create a common association of this alliance. Above the pickleweed zone, salt grass (*Distichlis spicata*) flats and alkali weed (*Cressa truxillensis*)–salt grass playas and sinks are common herbaceous alliances. A number of small shrubs such as seablite (*Suaeda esteroa, S. taxifolia*) and alkali heath (*Frankenia salina*) often occur. Less frequent species such as Parish's glasswort (*Arthrocnemum subterminale*), and Palmer's alkali heath (*Frankenia palmeri*) are also characteristic of more inland salt deserts far south into Mexico, underscoring the hotter, dryer setting of the southern California high salt marshes.

Freshwater Marshes

Upstream from salt marshes, freshwater marshes occur where slow-moving fresh water allows the establishment of emergent vegetation. Historically, these habitats were not common in coastal southern California because summer drought precluded the accumulation of the standing water required by marsh vegetation, and today most of our larger watercourses are channelized. Some estuaries today support freshwater marsh vegetation because agricultural runoff has added summer water to the drainages. The best examples of freshwater marsh vegetation occur in upper Newport Bay. Big Canyon at its mouth has a restored freshwater marsh fed by suburban runoff, and the University of California at Irvine supports the San Joaquin Marsh as part of the University of California Natural Reserve System. Adjacent to the university facility is a system of ponds, supported by the Irvine Ranch Water District, known as the San Joaquin Wildlife Sanctuary. Small isolated marsh areas are found at various ponds in the Santa Monica Mountains and at coastal creeks in San Diego County.

As salt concentrations are diluted with fresh water, vegetation changes from the upper high marsh halophytes to brackish water tolerant plants, and finally to fresh water species. Brackish water vegetation such as American bulrush (*Schoenoplectus americanus*) marshes and salt marsh bulrush (*Bolboschoenus maritimus*) marshes occur in typically narrow zones where the salt marsh edge is infused with fresh water from seeps, streams, or river mouths. Occasional brackish meadows of creeping rye grass (*Leymus triticoides*) and yerba mansa (*Anemopsis californica*) cover the ground in moist brackish soils. Once inland from the brackish water, vegetation is variously dominated by dense stands of cattail

Top: Yerba mansa stand at the Ballona Wetlands Ecological Reserve, Los Angeles County.

Right: Cattail stand and salt grass on Santa Rosa Island, Santa Barbara County.
Todd Keeler-Wolf photos

Above: Catalina ironwood forms an endemic tree alliance on the islands.
Todd Keeler-Wolf

Inset: Close-up of ironwood in bloom.
Allan Schoenherr

(*Typha* [*angustifolia, domingensis, latifolia*]) marsh and large bulrushes, including hardstem and California bulrush (*Schoenoplectus acutus* and *S. californicus*).

Occasional freshwater wetlands exist well inland from the coast. Some of the most interesting occur along fault scarps. These rare settings such as Etiwanda Marsh along the Cucamonga Fault have a diversity of wetland species in a patchwork of wetland alliances.

The Channel Islands

The eight Channel Islands are essentially mountaintops projecting above the sea, thrust up along a series of faults whose northward motion has carried them hundreds of miles from their points of origin. The larger islands remained above sea level during the period of translocations so some of the plant species found here have their origins much farther south on the mainland.

The four northern Channel Islands—San Miguel, Santa Rosa, Santa Cruz, and Anacapa— are aligned parallel to the mainland in such a way as to appear to be an extension out to sea of the Santa Monica Mountains. In fact during a period of lowered sea level 17,000 to 18,000 years ago they were connected in a land mass known as Santa Rosae (Schoenherr, et al. 1999). Even at that time a channel 7 km wide remained, separating them from the mainland. Thus, on all eight of the islands the present-day inhabitants had to arrive by some means of over-water transport or were carried on ocean currents to their present locations from a region farther south like passengers on a ship.

The four southern islands—San Nicolas, Santa Barbara, Santa Catalina, and San Clemente—form a diamond pattern in the sea. All of the islands today are managed in some form of protected status. The two outermost islands, San Clemente and San Nicolas, belong to the U.S. Navy. The four northern Channel Islands and tiny

Santa Barbara are managed as Channel Islands National Park, although 90% of western Santa Cruz is managed by The Nature Conservancy, and the U.S. Navy has retained the option of future use on northernmost San Miguel Island. Parts of Santa Catalina Island remain in private hands, but the Catalina Conservancy manages 86% of the island. The Santa Catalina Island Company owns most of the land around Avalon and the Two Harbors area, as well as a horse ranch (El Rancho Escondido) in the center of the island.

These islands are important examples of evolutionary laboratories. Long periods of isolation have resulted in significant endemism and many important variations in what, for the most part, are vegetation types characteristic of the mainland. Isolation tends to favor endemics because immigrants tend to keep evolving after their arrival. Species isolated on islands also can remain as relicts or remnants of populations that formerly enjoyed widespread distribution on the mainland. Numerous species of plants and some animals are limited in distribution to one or more of the islands. Catalina ironwoods are found in small groves on several islands, but they are absent from the mainland today. About 12,000 years ago, ironwoods were probably widespread on the mainland, but as the climate warmed and dried they were able to persist on a few steep hillsides on islands where, in the absence of competition, they survived in a cool, foggy climate that may resemble that of the past. Many island endemics are varieties or subspecies of mainland plants; only the ironwoods represent a distinctive genus. There are two dissimilar subspecies, however. Santa Cruz Island ironwood (*Lyonothamnus floribundus* ssp. *asplenii-folius*) is found on San Clemente, Santa Cruz, and Santa Rosa Islands. Santa Catalina Island ironwood (*L. f.* ssp. *floribundus*) is only found on Santa Catalina.

Larger islands are not only larger targets for dispersal, but they tend to have a greater diversity of habitats, including a diversity of soils (Schoenherr 1992). Santa Cruz and Santa Catalina Islands feature the greatest number of native taxa—both are approximately 30 km from the mainland. Both islands have a mountainous landscape: Santa Cruz reaches 753 m in elevation and Santa Catalina reaches 648 m. Santa Rosa is third in number of native species, but is farther from the mainland (44 km) and not as high (484 m). The two smallest islands, Anacapa and Santa Barbara, are about the same size—less than three square km—and have the fewest native

Endemic buckwheats, live-forevers, and California desert-thorn (Lycium californicum), *among others, cling to the salt-sprayed cliffs along Anacapa Island.*
Kendra Sikes

Top: Some vegetation types are restricted to specific soil types. On Santa Rosa Island, island chaparral is restricted to well-cemented sandstones.
TODD KEELER-WOLF

Above: Bright-green dudleya on Santa Catalina Island.
ALLAN SCHOENHERR

taxa. Anacapa Island has about twice the number of taxa (190) and it is only 20 km from the mainland, whereas Santa Barbara is 61 km away.

Most of the plant communities of coastal southern California are also present on the Channel Islands. Terrestrial vegetation consists of species associated with coastal sage scrub and coastal strands and dunes, although the larger islands also have chaparral, pine, and oak woodlands. While these general vegetation formations are similar, the details in species composition and evolutionary history are different. In some cases such as with the unusual endemic Catalina ironwoods and island oaks, the plants are strongly divergent from any species today on the mainland and represent ecological relicts. On the other hand, there are many unique species or subspecies of plants on the islands that have clear relatives on the mainland and represent recent diversification.

Rapid evolution into distinctive subgroups during isolation is called adaptive radiation, a characteristic of several groups of plants on the Channel Islands. For example, in the buckwheat family (Polygonaceae) there are at least twenty endemic species and subspecies. The largest number of endemics in a single genus occurs in the live-forevers (*Dudleya* spp.). Although taxonomic status of these cliff-dwelling leaf succulents is in a state of flux, at least nine species have distributions limited to islands and several others are shared with the mainland. For example, bright-green dudleya (*Dudleya virens*) also occurs on the Palos Verdes peninsula, its distribution being similar to Catalina crossosoma (*Crossosoma californicum*). Other groups that include a significant number of endemics are the manzanitas (*Arctostaphylos* spp.), bedstraws

(*Galium* spp.), desert dandelions (*Malacothrix* spp.), phacelias (*Phacelia* spp.), and deer weeds (*Lotus* spp.).

Two widespread live oaks of the mainland are also found on some of the islands: coast live oak (*Quercus agrifolia*) is found on Santa Cruz and Santa Rosa Islands, and the most widely distributed oak in California, canyon live oak (*Q. chrysolepis*), is found on Santa Catalina and Santa Cruz. A widespread island endemic, island oak is found on north-facing slopes and canyons of Anacapa, Santa Cruz, Santa Rosa, San Clemente, and Santa Catalina, as well as Guadalupe Island—an oceanic island far to the south off Baja California.

Of the scrub oaks, island scrub oak (*Q. pacifica*) is endemic to Santa Catalina, Santa Cruz, and Santa Rosa Islands where it is locally common and grows to large size, a form of gigantism. On Santa Catalina, island scrub oak forms an alliance of tree-sized plants that resembles oak woodlands of the mainland. Resembling a shrub form of interior live oak (*Q. wislizeni*), Santa Cruz Island oak (Q. *parvula* var. *parvula*) is found on Santa Cruz Island, and it also occurs in Santa Barbara County on the mainland.

Of the deciduous oaks, blue oak (*Q. douglasii*) and valley oak (*Q. lobata*) are found in a few woodland groves on Santa Catalina and Santa Cruz Islands. Another distinctive deciduous oak, MacDonald oak (*Quercus × macdonaldii*), is considered a hybrid between valley oak and island scrub oak. It only occurs on Santa Catalina, Santa Cruz, and Santa Rosa.

Gigantism is a phenomenon documented in animals and plants that have evolved on islands. Among the buckwheats, usually components of coastal sage scrub or island bluff scrub, there are large endemic species on every island. Santa Cruz Island buckwheat is on

Above: A large island oak on Santa Rosa Island.
Kendra Sikes

Left: Island scrub-jay (Aphelocoma insularis) *perched on lemonade berry.*
John Sterling

three of the northern Channel Islands. Santa Barbara, San Clemente, and Santa Catalina each has a variety of St. Catherine's lace (*Eriogonum giganteum*). Three varieties of island buckwheat (*Eriogonum grande*) are distributed among all eight of the islands. Giant coreopsis, a large member of the sunflower family (Asteraceae), may reach 2 m in height. Both succulent and drought-deciduous, it is well adapted for dry sites such as cliff faces and sandy soil. When covered with its spectacular yellow flowers, giant coreopsis is conspicuous on every island except San Clemente. It also occurs at a few localities on cliffs along the mainland coast. For many people this species has become a symbol of the Channel Islands.

Native pines are not widely distributed on the Channel Islands, but their distribution is interesting. There are two native pine species on the local islands and a third, Monterey pine (*Pinus radiata*), is found on Guadalupe and Cedros Islands off of Baja California. Santa Rosa Island pine is an endemic subspecies of Torrey pine (*Pinus torreyana* ssp. *insularis*). Bishop pine (*P. muricata*), found at spotty coastal locations on the mainland, occurs on Santa Cruz and Santa Rosa Islands.

Island Plant Communities

Most of the plant communities of coastal southern California are present on the Channel Islands. Most of the

Top: Island scrub oak becomes tree-like in size on Santa Catalina Island, an example of gigantism.

Center: St. Catherine's lace, an endemic variety of buckwheat on Santa Catalina Island, can reach heights of over a meter.

Left: Bishop pines are natives found on Santa Rosa Island.
Allan Schoenherr photos

Above: Giant coreopsis scrub on a coastal bluff along San Miguel Island.
Julie M. Evens

Left: Torrey pines on Santa Rosa Island.
Todd Keeler-Wolf

Above: Menzies' golden bush and San Miguel milkvetch on Santa Cruz Island.
TODD KEELER-WOLF

terrestrial vegetation consists of species associated with coastal sage scrub, although the larger islands also have chaparral and oak woodland. Coastal strand and dune vegetation still occur on some of the islands, and freshwater marsh occurs at a few localities on the larger islands. The southern islands have small stands of maritime succulent scrub, and island bluff scrub is well established on the islands.

While coastal sage scrub is common, its species composition varies among the islands. For example, coyote brush is the dominant plant of coastal scrub on the northern Channel Islands. Menzies' golden bush dominates on much of San Nicolas Island and it is important on the northern Channel Islands. Gum plant (*Grindelia stricta*), another yellow sunflower important on the northern Channel Islands, is common on San Miguel and Santa Rosa, but dominant on parts of Anacapa. Silver bush lupines are common on the northern Channel Islands and become dominant on portions of Santa Rosa, San Miguel, and San Nicolas (Junak 2008). Yellow bush lupine (*Lupinus arboreus*) is also present on San Miguel and Santa Rosa. Laurel sumac is absent from the northern Channel Islands and black sage is rare on San Clemente, San Nicolas and San Miguel. An endemic relative of black sage, Santa Rosa Island sage (*Salvia brandegeei*), which unlike any other mainland sage is "scentless," replaces it on Santa Rosa Island. California sagebrush is replaced by the endemic island sagebrush (*Artemisia*

Above: Silver bush lupine, gaining in dominance and stabilizing a sand dune just inland from the coast, on San Miguel Island. Julie M. Evens

Left: Island sagebrush is a dominant drought-deciduous component of coastal sage scrub on Santa Barbara, San Nicolas, and San Clemente Islands. Todd Keeler-Wolf

Top: View from Inspiration Point, Anacapa Island, with island buckwheat on its steep northerly slope.
TODD KEELER-WOLF

*Above: Orange-crowned warbler (*Oreothlypis celata sordida*) perched on island mallow.*
JOHN STERLING

nesiotica) on Santa Barbara, San Clemente, and San Nicolas Islands.

Grazing by introduced herbivores and rooting by pigs have reduced the cover and diversity of coastal scrub on most of the islands. As a result, off-shore islets, cliffs, sea bluffs, and cactus patches have become refugia for rare and endemic plants. This feature also has favored de facto preservation of the coastal bluff scrub community. Endemic Nevin's eriophyllum (*Eriophyllum nevinii*), of Santa Barbara, Santa Catalina, and San Clemente, and island tarweed (*Deinandra clementina*) have been preserved this way. Giant coreopsis, various live-forevers, the large buckwheats, and desert dandelions, many of them endemics, cling to these inaccessible sites. One of the most showy island shrubs, island mallow (*Lavatera assurgentiflora*) is a component of coastal bluff scrub. The northern form (*L. a.* ssp. *assurgentiflora*), which has become naturalized on the mainland and is available as a nursery plant, is native to Anacapa and San Miguel.

The coastal strand communities—beach, foredune, and dune scrub—are largely absent from the mainland due to serious disturbance, but they are well represented on most of the islands (Junak et al. 2007). Anacapa and Santa Barbara lack sandy beaches or dunes. Low growing, sprawling plants that tolerate winds, salty soil, and shifting sand dominate these habitats. Many of them are related to desert plants. Common among them are red sand verbena (*Abronia maritima*), beach bur-sage, beach evening primrose, and beach morning-glory. Non-native ice plants, including Hottentot fig (*Carpobrotus edulis*) and crystalline ice plant (*Mesembryanthemum crystallinum*), also colonize these beaches. The windiest islands—San Miguel, Santa Rosa, Santa Cruz, San Nicolas, and San Clemente—have regions of active dunes sometimes sweeping across entire islands. Yellow bush lupine and silver bush lupine totally dominate the community at some locations. Common associates include Menzies' golden bush and San Miguel milkvetch (*Astragalus miguelensis*).

Maritime succulent scrub is a community that appears only on San Clemente, San Nicolas, and in a much restricted form on the east end of Santa Catalina. It is characterized by many succulent species, with golden cereus a conspicuous dominant.

Habitat Degradation and Recovery

Influence of humans on the Channel Islands has been profound, including the impact of military occupation and the well-documented history of grazing and its effect on native plants. Evolution in the absence of large herbivores means that many native island species lack bitter chemicals or unpleasant odors that discourage introduced grazing animals. Probably the most insidious impact has been the replacement of native species by non-native animals and plants. Non-native grass species outnumber native ones and cover more land: on Santa Cruz Island, 28 species of native grasses are now outnumbered by 42 non-native species. On Santa Rosa Island, grassland covers 67% of the island's surface. These grasslands are composed primarily of oats (*Avena* spp.), bromes (*Bromus* spp.), barleys (*Hordeum* spp.), and rye grasses (*Lolium* spp.). For the most part these species were introduced in the fur, food, and droppings of domestic

Top: Red sand verbena and beach bur-sage, a foredune association, on San Miguel Island.

Above: Plants clinging to this cliff on San Miguel Island include red-flowered buckwheat, live-forever, and giant coreopsis.

Todd Keeler-Wolf photos

livestock. Furthermore, many of the dominant plants are introduced ornamental species, particularly in the vicinity of Avalon, Santa Catalina Island. French broom (*Genista monspessulana*) with its bright yellow flowers is now a dominant plant in coastal sage scrub there. Tasmanian blue gum (*Eucalyptus globulus*) is the most conspicuous tree on most of the islands.

*Above: San Miguel milkvetch, dunedelion (*Malacothrix incana*), and island paintbrush (*Castilleja lanata *ssp.* hololeuca*) on San Miguel Island.* Todd Keeler-Wolf

Left: Crystalline ice plant, a non-native species, has colonized beaches and bluffs on all the islands.
Julie M. Evens

Opposite top: Recovered vegetation on the dunes of San Nicholas Island.

Opposite bottom: French broom and Tasmanium blue gum are conspicuous non-native plants on Santa Catalina Island near Avalon.
Allan Schoenherr photos

Fortunately, the agencies which manage all eight of the Channel Islands are, to a large degree, motivated by preservation and restoration. Most of the seriously degraded lands are showing remarkable signs of recovery. On the other hand, vegetation on southern California's coastal mainland will continue to be influenced by human activities. While further development is limited by protection on public lands, other factors such as invasive species, air pollution, changes in fire dynamics, pathogens, and climate change will remain active threats to native vegetation.

Allan A. Schoenherr

CHAPTER TWELVE

Southern California Mountains and Valleys

California's South Coast Ranges encompass two major mountain ranges: the east-west trending Transverse Ranges and the north-south oriented Peninsular Ranges. The Transverse Ranges are situated between the coastal plains of Santa Barbara, Ventura, Los Angeles, and San Bernardino to the south, and the inland San Joaquin Valley or Mojave Desert to the north. The Peninsular Ranges are situated between the ocean coast to the west and the inland Mojave Desert to the east.

The Transverse Ranges are best known for the spectacular "wall" of mountains that forms the precipitous boundary north and east of the Los Angeles Basin, but include an equally large area of lower mountains west to Point Conception. They consist of three distinct mountainous regions. Except for Mount Pinos, the Western Transverse Ranges are composed almost entirely of sedimentary sandstones. The near-coast Santa Ynez Mountains rise steeply behind Santa Barbara and Ojai. The more inland San Rafael and Sierra Madre Mountains rise just north of the Santa Ynez Valley. The Sierra Madre Mountains mark the region's northern boundary. The Santa Ynez, San Rafael, and Sierra Madre Mountains reach a maximum height of around 1,980 m, mostly blanketed in chaparral, but not high enough to support significant montane conifer forests. Montane conifer forests in the San Rafael and Sierra Madre Mountains are isolated sky islands on steep, north-facing slopes. In contrast, the higher, more

Above: The San Gabriel Mountains rise precipitously above the Los Angeles Basin and are part of the Transverse Ranges.
Michael E. Kauffmann

Opposite: The Peninsular Ranges are situated between the coast and the Mojave Desert, shown here in Carrizo Gorge Wilderness, north of Interstate 8 and east of San Diego.
Robert Wick

Top: Recovering burned chaparral in the 2009 Station Fire, Los Angeles County.
Carole Scurlock

Above: Xeric chaparral dominated by chamise on Otay Mountain, San Diego County.
Thomas Oberbauer

inland Mount Pinos and Sawmill Mountain areas support extensive montane conifer forests. The Castaic Mountains, east of Mount Pinos and Interstate 5, are the easternmost outposts and the northern boundary of the Western Transverse Ranges.

East of the Western Transverse Ranges, the precipitous San Gabriel Mountains tower above the densely populated Los Angeles Basin. The San Gabriel Mountains are primarily composed of ancient crystalline rocks (e.g. quartzite, shist, marble). Further east, the San Bernardino Mountains rise above the cities of San Bernardino, Yucaipa and Banning. The eastern terminus of the Transverse Ranges, the San Bernardino Mountains, are also composed of a complex mixture of crystalline (granites, marbles, quartzites) as well as sedimentary rocks. Both the San Gabriel and San Bernardino Mountains reach heights of 3,500 m and support extensive montane and even subalpine forests.

The island-like San Jacinto Mountains mark the northern end of the Peninsular Ranges, several complex subunits stretching south well into Baja California. To the west of the San Jacinto Mountains the low, coastal Santa Ana Mountains include the Elsinore and Santa Margarita Mountains. The coastal plains west of these mountains are densely populated all the way to the coast. Directly south of the San Jacinto Mountains, the more diffuse San Diego Ranges extend almost to the Mexican border. The San Diego Ranges include the southeastern Cuyamaca Mountains and Laguna Mountains, which demarcate the eastern boundary of the region at the Sonoran–Colorado desert's edge. Much of the rest of the San Diego Ranges encompass a variety of mountains, foothills, valleys, and plateaus.

Vegetation types in this region can be described according to the major landscapes in southern California (Stephenson and Calcarone 1999). Coastal foothills and lower montane landscapes are ocean-facing and range from elevations of 90 to 1,525 m. Chaparral dominates both landscapes. Grasslands, oak woodlands, and coastal sage scrub are prevalent at the lower elevations (less than 900 m) while bigcone Douglas fir (*Pseudotsuga macrocarpa*) and Coulter pine (*Pinus coulteri*) forests are best represented from 900 to 1,500 m. The montane conifer landscape occupies elevations from 1,500 to 2,600 m and is dominated by pine (*Pinus* spp.) and fir (*Abies* spp.) forests. The subalpine–alpine landscape occurs above 2,600 m. The desert montane landscape ranges from 900 to 2,100 m but is arid and situated inland of the other landscapes.

Top: The San Jacinto Mountains, along a steep fault block of granitic rocks ascending 3,000 m, show a transition from lower elevation chaparral with bigberry manzanita, redshank, and chamise to higher elevation forests with California black oak, canyon live oak, and pines.
Julie M. Evens

Above: Chamise is ubiquitous in chaparral and often forms extensive pure stands.
Thomas Oberbauer

Chaparral

Beginning in the Santa Ynez Mountains at Gaviota Pass west of Santa Barbara—and extending to the Otay Mountains at the Mexican border south of San Diego—chaparral is the most prevalent vegetation in the South Coast Ranges and covers nearly 35% of the area. Precipitation in chaparral-covered landscapes falls almost entirely as rain and averages between 380 mm and 750 mm. Winter temperatures generally remain above freezing.

All plant alliances in the South Coast Ranges are to varying degrees influenced by fire, but fire has a particularly prominent role in the ecology of chaparral. Winter and spring precipitation keeps plant moisture high through the spring and early summer and shrubs usually do not easily burn. By fall, as soil moisture is depleted, plant moisture drops quickly. Chaparral wildfires—especially those driven by hot, dry Santa Ana winds—erupt in the fall, spread quickly, burn at high intensities, and have the potential to reach thousands of hectares in size. The frequency of fires has profound effects on the plant species composition and geographic extent of chaparral.

These chaparral shrublands are diverse in shrub species. Chamise (*Adenostoma fasciculatum*) occurs nearly everywhere while other shrubs in the

Top: Bigberry manzanita and chamise are common xeric chaparral species.
JEFF BISBEE

Above: Hoary leaf ceanothus is a xeric chaparral alliance.
JULIE M. EVENS

genera *Arctostaphylos* and *Ceanothus* range from widespread in their distributions, to local and rare. More than 60 chaparral alliances have been cataloged to date and many more associations will be added as classification efforts continue.

The chaparral alliances can be grouped into two major types that internally share important life history and ecological characteristics. The xeric (warm and dry) group of alliances is composed of species that inhabit drier environmental settings, such as south-facing slopes that intercept higher amounts of solar radiation year around. Xeric alliances such as bigberry manzanita (*Arctostaphylos glauca*) chaparral and wedge leaf ceanothus (*Ceanothus cuneatus*) chaparral are dominated by species which are drought-tolerant both as seedlings and adults. Nearly all of these species are "obligate seeders," fires routinely kill adult shrubs. Instead of resprouting, the new generation of shrubs emerges entirely from dormant seeds that accumulate in the soil between fires. Heat and even smoke from the fire is required to break the hard seed coats of many of these shrubs and enable seeds to soak up winter rains and germinate. As a result, nearly all seedlings appear in the first spring after a fire. They grow quickly in height and begin producing seeds at an early age, which replenishes the soil seed bank before the next inevitable fire occurs. Most species in xeric chaparral disperse seeds passively beneath the parent plants and animal dispersal of seeds is uncommon. There are almost as many ceanothus alliances in the South Coast Ranges as there are species and the majority are obligate seeders in xeric alliances. Common xeric alliances include hoary leaf ceanothus (*Ceanothus crassifolius*), wedge leaf ceanothus, bigpod ceanothus (*C. megacarpus*), and cup leaf ceanothus (*C. greggii*) (Gordon and White 1994; Borchert et al. 2004; Klein and Evens 2005; Keeley and Davis 2007).

The second set of shrub species form alliances in the mesic chaparral group. "Mesic" refers to the shaded, relatively moist north-facing slopes where annual solar radiation is relatively low. Typical mesic alliances include toyon (*Heteromeles arbutifolia*) chaparral, scrub oak (*Quercus berberidifolia*) chaparral, and California coffee berry (*Frangula californica*) scrub. Mesic plants are almost always "obligate resprouters" in contrast to the xeric obligate seeders. They are not drought-tolerant like the xeric obligate seeders but avoid wilting in the long, rainless summers by virtue of extensive root systems that tap into deep soil moisture. Instead of regenerating entirely from seeds after fire, mature plants resprout from underground living organs (burls, corms, bulbs, etc.). In sharp contrast to obligate seeders, seedlings of resprouters are uncommon, or even absent, after fire. Moreover, resprouter seeds lack dormancy and do not accumulate in the soil as a dormant seed bank but instead germinate soon after ripening. Apparently, seedlings in this group germinate and grow best in the deep shade and litter of old-aged chaparral (Keeley and Davis 2007).

Top: Xeric chaparral dominated by bigpod ceanothus in the Santa Ynez Mountains. Jeff Bisbee

Middle: Mesic chaparral dominated by scrub oak in the Santa Ana Mountains.

Above: Laurel sumac, a facultative seeder, in the Santa Monica Mountains. Todd Keeler-Wolf photos

Top: Bigcone Douglas fir is endemic to southern California and typically occurs in highly fragmented stands embedded in chaparral.
Michael E. Kauffmann

Above: Greenbark ceanothus is a mesic chaparral alliance.
Julie M. Evens

Another group of species is called "facultative seeders." These are shrubs that both resprout and establish abundant seedlings after fire. However, these species do not fit neatly into the xeric or mesic groups based on life history characteristics. For example, chamise is the most common and widespread facultative seeder. The chamise alliance is a member of the xeric group even though it is not an obligate seeder. On the other hand, facultative seeders such as greenbark ceanothus (*C. spinosus*) and laurel sumac (*Malosma laurina*) are members of the mesic alliances.

Bigcone Douglas Fir and Canyon Live Oak Forests

Bigcone Douglas fir is an unusual southern California endemic. Its name suggests an unusually large cone, but in reality, its cones are closer in size to the cone of Jeffrey pine (*Pinus jeffreyi*) and only somewhat larger than the ponderosa pine cone. Nevertheless, its seeds and cones are large compared to its widespread northern cousin Douglas fir (*Pseudotsuga menziesii*) with its comparatively small cones and more numerous small seeds.

Bigcone Douglas fir forest alliance ranges from the southern Santa Lucia Mountains to Otay Mountain near the Mexican border, but does not extend into Mexico. Unlike the extensive unbroken forests of Jeffrey pine in the San Bernardino Mountains, on Mount Pinos or in the Laguna Mountains, bigcone Douglas fir forest has a highly fragmented, archipelago-like distribution composed of thousands of individual stands that rarely exceed 2–5 ha in size. These patches of candelabra-shaped trees inhabit a recognizable environmental setting. Trees cling to the steepest slopes tucked away in canyons, coves, and ravines and can even grow directly from sheer rock walls. Many trees and shrub species co-occur with bigcone Douglas fir over its range but its most predictable associate is canyon live oak (*Quercus chrysolepis*) (Minnich 2007).

Fire is the principal reason the bigcone Douglas fir forest alliance occupies these precipitous settings. Typically, forests are isolated groves of trees surrounded by a continuous cover of chaparral. Because chaparral invariably burns in crown fires, it poses a direct threat to bigcone Douglas

fir forests. Where bigcone Douglas fir grows, however, understory shrub and herbaceous vegetation typically is sparse. Considerable bare ground prevails on the steep, erosive slopes and rock outcrops. At first fires burn intensely through the bordering chaparral but when they encounter these forest pockets, fire behavior can quickly change from high to moderate intensity, or even to low intensity surface fires (Minnich 2007).

Mature bigcone Douglas fir trees have thick bark and the lower canopy branches grow well above the ground. As a result, fire may char the well-protected lower trunks but not spread upward into the canopy. If fire does carry up the tree bole and partially burns the foliage, bigcone Douglas fir has the unusual capacity to resprout from dormant buds hidden beneath the bark of the branches. In fact, it is only one of three conifers in the world capable of resprouting in response to physical injury. The capacity to canopy sprout, however, is limited. High-intensity fires frequently burn through most of the canopy and kill the tree. Previously killed trees turn grayish-white over time and are intermixed with live trees less affected by the fire. Firs on the edge of the stands nearest the chaparral usually succumb to fire while those in the interior survive. Trees that crown sprout after fire have a distinctive pipe-cleaner appearance. Occasionally, entire fir pockets are killed—leaving only the resprouting canyon live oak to reclaim the site.

The canyon live oak forest alliance grows on the steep, arid, rocky slopes of lower montane forests. Although it often associates with bigcone Douglas fir, it

Top: Bigcone Douglas fir frequently associates with canyon live oak.
THOMAS OBERBAUER

Above: A multi-stemmed oak provides a secure base for the stick nests of big-eared woodrats.
JONAH EVANS

Top: Cuyamaca cypress forms small groves on Cuyamaca Peak, San Diego County.
MICHAEL E. KAUFFMANN

Above: Tecate cypress grows exclusively on poorly developed soils with low nutrient levels; Otay Mountain, San Diego County.
THOMAS OBERBAUER

also occurs as dense forest patches and ranges to higher elevations where it locally becomes a constituent of other conifer forests. Canyon live oak is well adapted to fire-prone, arid settings because it taps deep water sources with its extensive root system. It is easily top-killed by fire but resprouts prolifically from the bole or root crown. Fire-caused resprouting creates multi-stemmed trees that provide a secure base for the stick nests of big-eared woodrats (*Neotoma macrotis*) that seasonally consume acorn crops.

Cypresses

Three cypress species occur in southern California: Sargent cypress (*Hesperocyparis sargentii*), Cuyamaca cypress (*H. stephensonii*), and Tecate cypress (*H. forbesii*). All are rare and distributed in small, widely scattered populations. The southernmost stand of Sargent cypress woodland is in the San Rafael Mountains. Cuyamaca cypress stands, the most restricted southern California cypress, exist as a collection of several groves in the Cuyamaca Peak/King Creek area of the Cuyamaca Mountains covering only about 90 ha. Tecate cypress stands grow on alkaline clay soils derived from nutrient-deficient metavolcanic and gabbro rocks on Otay Mountain, Tecate Peak, and Guatay Mountain in San Diego County and on

Paleocene sandstone in Gypsum Canyon in the Santa Ana Mountains of Orange County (Minnich 2007).

The cypresses all grow within or in close proximity to highly flammable chaparral and nearly always succumb to periodic wildfires. To persist in an environment of high-intensity fires, cypress species are serotinous. They develop a canopy seed bank of small closed cones containing tiny seeds. Cones accumulate by the thousands on tree branches in the years between fires. Although many open as they dry, many more remain closed, sometimes for decades. Heat from the next inevitable fire opens the cones, releasing millions of seeds to the fertile ash bed. Seeds germinate after winter rains, often forming carpets of seedlings by the following spring.

All three cypress species are vulnerable to what fire ecologists term "immaturity risk." Seedlings that appear after fire are in a race to accumulate a large collection of cones to re-seed after the next fire. For example, Tecate cypress begins to produce cones at about ten years of age. Cones accumulate rapidly over the next several decades and achieve peak cone production when trees reach about 50 years. The stand will only regenerate when trees produce a critical number of closed cones. Fires that burn before the cone threshold is attained may result in poor seedling establishment, which threatens the future of the stand.

As human-caused fires have become the most prevalent source of wildfires in chaparral, the interval between fires has become increasingly, and unnaturally, short. One consequence of this shortened interval is that species that regenerate entirely from seed run the risk of being eliminated, especially if they do not produce enough seeds to re-establish the stand after the next fire. Moreover, serotinous species such as Tecate cypress, which requires an

Above: Tecate cypress seedlings and saplings recruit after a wild-fire in the northern Santa Ana Mountains, Orange County.
TODD KEELER-WOLF

Below: Mixed chaparral dominant in areas with repeat burning of Tecate cypress on Otay Mountain include chamise, southern mountain misery (Chamaebatia australis), *Otay ceanothus* (Ceanothus × otayensis), *and hairy leaf ceanothus* (C. oliganthus).
MICHAEL GOLDEN

extended period of time (up to 40 years) to produce a cone bank of sufficient size, are at even greater risk of extinction (de Gouvenain and Ansary 2006).

Engelmann and Coast Live Oak Woodlands

Of California's oak species, Engelmann oak (*Quercus engelmannii*) has the smallest range covering only 31,500 ha. It is distributed from the San Gabriel Mountains to Baja California—but 93% of its highly fragmented distribution is concentrated between Palomar Mountain and Cuyamaca Peak in San Diego County. Engelmann oak woodland, the southern California equivalent of blue oak woodland, has an understory dominated by grasses and herbs but rarely occurs in single-species woodlands. Instead, coast live oak (*Q. agrifolia*) is often co-dominant. Organ Valley Research Natural Area in the Cleveland National Forest and Santa Rosa Plateau Ecological Reserve, protected by The Nature Conservancy, are excellent locations to see these woodlands in varying tree densities.

Montane Conifer Landscape

Yellow pine forests and mixed conifer forests are the two major vegetation types comprising the montane conifer forests broadly spanning elevations of 1,500–2,300 m in the South Coast

Top: Mixed conifer forest with ponderosa pine, incense cedar, and California black oak on the lower slope, and Coulter pine with California black oak or canyon live oak and Pringle manzanita on the mid-upper slope, in the San Jacinto Mountains.
Julie M. Evens

Middle: Engelmann oaks, as savannas and woodlands, have grass and herb understories at Santa Ysabel Preserves.

Left: Jeffrey pine, white fir, oak, and incense cedar in a mixed conifer forest on Palomar Mountain, San Diego County.
Thomas Oberbauer photos

Ranges. Where precipitation increases (635–1,270 mm) above the lower montane conifer zone and average minimum temperatures decline to 0–5°C, yellow pine forest composed of ponderosa pine (*Pinus ponderosa*) and Jeffrey pine dominate, with each pine forming its own alliance (Minnich 2007). Spatially intermixed with yellow pine alliances are mixed conifer forests. "Mixed conifer" is a catch-all term used to describe forests composed of an array of conifer species, none of which consistently dominates the species mix. However, as quantitative data amasses for these forests, vegetation ecologists are finding that mixed conifer forests are made up of a number of different alliances that more accurately capture the compositional variation within this forest type, as well as subtle differences in soil moisture, texture, temperature, and depth associated with each.

Ponderosa pine forests are encountered at the lowest elevations of the montane conifer zone, from 1,625–2,100 m which is typically where snowfall becomes the significant form of precipitation and winter temperatures predictably fall below freezing. Although ponderosa pine dominates this alliance,

Top: Upper montane mountain whitethorn shrubland growing next to Jeffrey pine woodlands. Richard Spjut

Above: Coulter pine and bigcone Douglas fir are components of a low-elevation mixed conifer forest on the eastern slope of Palomar Mountain, San Diego County. Thomas Oberbauer

conifers and hardwoods such as incense cedar (*Calocedrus decurrens*), sugar pine (*Pinus lambertiana*), white fir (*Abies concolor*), California black oak (*Quercus kelloggii*), and canyon live oak are frequently present. Common understory shrub species include California coffee berry, deer brush (*Ceanothus integerrimus*) and Pringle manzanita (*Arctostaphylos pringlei*). Above the ponderosa pine belt, mixed conifer and Jeffrey pine forests become increasingly abundant (Minnich 2007).

Mixed conifer forests develop best on relatively moist, shaded north-facing slopes—although species composition can change markedly with both elevation and increasing aridity. In the South Coast Ranges white fir, sugar pine, and incense cedar typically co-mingle with Jeffrey pine. The consistent presence of white fir and sugar pine along with other less constant conifers is recognized as the white fir–sugar pine forest alliance. On the steepest slopes above 2,200 m the white fir forest alliance predictably co-occurs with mixed conifer forests. In the western San Gabriel Mountains Coulter pine is also abundant in mixed conifer forests, while in the San Bernardino Mountains bigcone Douglas fir locally mixes with other conifer species. Black oak, a common understory hardwood in mixed conifer forests, may also occur as a single species in woodlands or forests.

With increasing elevation, conditions become increasingly colder and drier. In more inland mountain settings, Jeffrey pine forests dominate south-facing slopes and flat areas between 2,100 and 2,700 m while mixed coni-

*Top: California juniper (*Juniperus californica*) woodland in San Felipe Valley, a more arid area of eastern San Diego County.*

Left: Mixed conifer, white fir–sugar pine forests tend to occur on more sheltered exposures where winter snowfall accumulates.
Thomas Oberbauer photos

fer forests are increasingly restricted to north-facing aspects. Common shrub species in Jeffrey pine forests are mountain whitethorn (*Ceanothus cordulatus*) and green leaf manzanita (*Arctostaphylos patula*). At higher elevations pine forests often give way to timberland chaparral composed of the latter two shrubs plus bush chinquapin (*Chrysolepis sempervirens*). Over large areas of nearly all the major mountain ranges the Jeffrey pine forest alliance blankets the landscape until it eventually transitions to more arid singleleaf pinyon (*Pinus monophylla*) woodlands or desert chaparral. On these desert margin forests several species co-mingle depending on slope exposure. These include singleleaf pinyon and junipers (*Juniperus occidentalis, J. osteosperma*, and *J. californica*) along with Coulter pine, canyon live oak, and bigcone Douglas fir (Minnich 2007).

Subalpine–alpine Landscape

Subalpine forests are patchy in distribution at the highest elevations (greater than 2,400 m) of the San Gabriel Mountains, San Bernardino Mountains, and San Jacinto Mountains. Lodgepole pine (*Pinus contorta* ssp. *murrayana*) forests mostly occur between elevations of 2,400–2,700 m and are represented in the San Gabriel, San Bernardino, and San Jacinto Mountains. Mountaintops receive substantial precipitation in the form of snow persisting well into the summer months. As temperatures remain low, the growing season is correspondingly short. At the highest elevations, tree stature decreases where icy winds pummel the peaks and ridges. Here pines decrease in height and become gnarled and twisted, forming low hedge-like krummholz. In the San Bernardino Mountains lodgepole pine forests may be dense and even-aged on lower north-facing slopes while forests in the San Jacinto Mountains are multi-aged, the result of low-intensity, lightning-caused fires and occasional stand-replacing fires. Above 3,000 m, limber pine (*Pinus flexilis*) forms open woodlands and may intermix with other conifers such as lodgepole pine, western juniper (*Juniperus occidentalis*), and singleleaf pinyon in open forests. San Gorgonio Mountain is probably the best location in southern California to view limber pine in all its forms, ranging from tall veterans in sheltered basins to stunt-

Top: Open limber pine woodlands above 3,000 m on Mount Pinos, Ventura County.
Michael E. Kauffmann

Above: Clark's nutcrackers cache enormous numbers of pine seeds and are vital to the dispersal of limber pine.
Claus Holzapfel

Top: Singleleaf pinyon pine forest with California juniper above Sespe Creek in the Los Padres National Forest, Ventura County. Michael E. Kauffmann

Above: Singleleaf pinyon pine woodland with big sagebrush in the San Bernardino National Forest, San Bernardino County. Steve Perkins

ed krummholz individuals on exposed, windswept ridges (Minnich 2007).

One of the most intriguing aspects of limber pine ecology is its dependence on the Clark's nutcracker (*Nucifraga columbiana*), a jay-sized member of the crow family (Lanner and Vander Wall 1980). Nutcrackers form permanent pairs and occupy subalpine forests year-round if possible. This bird has evolved several distinctive morphological adaptations for exploiting limber pine and other five-needled pines. It can break open cones or extract seeds from narrow cracks in the scales of opening cones with its long and sharply pointed bill. Clark's nutcrackers are fast and strong fliers, allowing them to search for cone-bearing trees within and between mountaintops and, after collection, transport seed long distances—up to 23 km. An expandable sublingual pouch allows them to collect and transport up to 150 seeds at a time. During the fall seed harvest, Clark's nutcrackers bury seeds in groups of 1 to 15 seeds in the soil at depths of 1 to 3 cm. Most seeds are cached within a few meters of the source tree, but nutcrackers also transport seeds long distances to communal caching sites that are snow-free in the winter.

To aid in re-finding caches, nutcrackers have remarkably specialized brains. They remember the precise locations of thousands of individual seed caches for as long as a year, even when the caches are covered by snow. Fortunately for limber pines, the nutcracker's ability to re-find caches diminishes with time and the locations of many caches are forgotten. Un-retrieved caches may germinate into seedling clusters that represent the next generation of limber pines.

Desert Montane Landscape

In the South Coast Ranges four extensive areas of pinyon woodland lie in the arid rain shadows along the desert margins of the major mountain ranges. Woodlands range in elevations from 1,200–2,400 m where summers are hot (31–35°C) and dry, and winters are cold (20–30°C) but receive small amounts of precipitation (average 230 mm). Except for the Western Transverse Ranges, pinyon woodlands form the transition between montane forests at higher elevations and the Joshua tree (*Yucca brevifolia*) woodlands and creosote bush (*Larrea tridentata*) scrub at lower desert elevations.

The rounded, somewhat conical canopies of singleleaf pinyon trees impart a rather uniform, even monotonous appearance to these woodlands—but this homogeneity can mask a surprisingly rich mixture of shrub and herbaceous species, especially when compared to the vast pinyon woodlands covering the Great Basin. The westernmost singleleaf pinyon woodlands extend from the Cuyama Valley through Lockwood Valley (Mount Pinos). Lockwood Valley Road from Highway 54 to Interstate 5 provides a perfect transect through these woodlands. Singleleaf pinyon is the dominant tree, but Tucker oak (*Quercus john-tuckeri*) is common in the understory along with scattered California junipers (*Juniperus californica*).

Along the northern base of the San Gabriel Mountains, California juniper co-dominates with singleleaf pinyon. Tucker oak, bitter brush (*Purshia tridentata*), desert almond (*Prunus fasciculata*) and curl leaf mountain mahogany (*Cercocarpus ledifolius*), are locally abundant in the shrub understory. Pinyon woodlands on the leeward side of the San Bernardino Mountains have been described in greatest detail and highlight the complex interaction of these woodlands with geology. At elevations from 1,200–2,400 m, singleleaf pinyon woodlands form monotypic stands. Characteristic understory species include narrowleaf goldenbush (*Ericameria linearifolia*), bitter brush, Great Basin big sagebrush (*Artemisia tridentata* ssp. *tridentata*), and Mormon tea (*Ephedra viridis*). Within this same elevational range, but preferentially growing on nutrient-poor carbonate soils, singleleaf pinyon shares the overstory with Utah juniper (*Juniperus osteosperma*) (Neel 2000). Characteristic understory elements on carbonate are crested needlegrass (*Achnatherum coronatum*) and green rabbitbrush (*Chrysothamnus viscidiflorus*). At higher elevations of 1,900–2,750 m, Utah juniper gives way to western juniper (which grows both on and off carbonate soils) and curl leaf mountain mahogany commonly associates with western juniper.

Further to the east, on the desert-facing lower slopes of the San Jacinto

Top: Redshank chaparral along desert-transition slopes, near Warner Springs in eastern San Diego County.
TODD KEELER-WOLF

*Above: Parry pinyon with birch leaf mountain mahogany (*Cercocarpus montanus *var.* glaber*) and redshank chaparral along Thomas Mountain near Anza, Riverside County.*
JOANNA LEMLY

Above: Spectacular spring flowering on the Sawmill Pebble Plain, surrounded by Jeffrey pine forest. This 67 hectare area, recently acquired by San Bernardino Mountains Land Trust, is protected and managed in partnership with the Big Bear Valley Education Trust.
Ed Wallace

Below: Southern mountain buckwheat, an endemic and federally threatened plant, is found on the unique Pebble Plain habitat—a biologically rich yet highly threatened plant community.
Elizabeth Eichorn (SBNF)

Mountains, singleleaf pinyon associates with many of the above-mentioned shrub species except that chaparral elements like redshank (*Adenostoma sparsifolium*) enter the shrub mix. The fourth area of pinyon woodlands occurs in the rain shadow of the Laguna Mountains in the southern Peninsular Ranges where the Parry pinyon (*Pinus quadrifolia*) woodland occupies the desert escarpment.

Pebble Plains

In the northeastern San Bernardino Mountains, in the vicinity of Big Bear Lake, is the anomalous Pebble Plain habitat. These are not plains in the conventional sense, such as the Carrizo Plain in the Central Coast Ranges. Instead, "plains" describes a widely scattered collection of dry, treeless patches which, owing to their unusual soils, support a distinctive suite of plant species that are mostly missing from the Jeffrey pine, western juniper, and pinyon pine forests that surround them. In all, there are 14 Pebble Plain complexes covering 1,620 hectares. Each complex is a collection of island patches in close proximity to one another. However, within each complex, patch size and species composition are variable.

The combination of soils, climate, and geologic time has created an environment for species capable of inhabiting this harsh habitat, much in the same way a distinctive group of species has evolved to withstand the chemical imbalances of serpentinite soils. Pebble Plains' soils are deep clays (~50%) in a matrix of quartzite pebbles and gravels (~50%). Geologists speculate these soils were originally Mojave pavements but were pushed to high-mountain elevations during the uplift of the San Bernardino Mountains in the last five million years.

In freezing winter temperatures, the wet clays in Pebble Plains' soils expand and push the cobbles and gravels to the surface, causing a slow-motion churning that impedes tree and shrub seedling establishment. Indeed, conifer colonization has been remarkably

unsuccessful in most Pebble Plains. In the dry summer months, after spring soil moisture disappears, the exposed soils bake in the sun, unprotected by tree shade or forest litter. As a result, Pebble Plains' species are low growing, have succulent leaves, and possess highly reflective plant parts—morphological characters that enable them to persist in this harsh habitat. Plant species such as Bear Valley sandwort (*Arenaria ursina*) and southern mountain buckwheat (*Eriogonum kennedyi* var. *austromontanum*) are endemic to the Pebble Plains while another buckwheat variety, Kennedy's buckwheat (*Eriogonum kennedyi* var. *kennedyi*), and fuzzy rattails (*Ivesia argyrocoma*) are restricted to the Plains, but are not endemics. Many other species are associated with this habitat but are not considered Pebble Plains' indicator species. In fact, an unusually high number of federally listed and rare species occur in the plains. The two buckwheat varieties serve as hosts for two rare butterflies, the Coxey Meadow butterfly (*Euphilotes baueri vernalis*) and the Baldwin Lake blue butterfly (*Euphilotes enoptes* near *dammersi*). Southern mountain buckwheat is the primary host for the parasitic ash-gray paintbrush (*Castilleja cinerea*), one of three threatened plant species.

Top: Desert ceanothus (Ceanothus greggii *var.* perplexans) *mixed with Eastwood's manzanita* (Arctostaphylos glandulosa) *in the Laguna Mountains.*
Jeff Bisbee

Above: Ash-gray paintbrush and southern mountain buckwheat on North Baldwin Pebble Plain.
Deveree Kopp

Semi-desert chaparral with California buckwheat, white sage (Salvia apiana), *and valley cholla.*
Todd Keeler-Wolf

Semi-desert Chaparral

Singleleaf pinyon woodlands do not always form the transition from montane forests to desert scrub. In some locales of the San Gabriel Mountains and San Bernardino Mountains, desert chaparral, also known as semi-desert chaparral, replaces singleleaf pinyon as the transitional vegetation. Compared to coastal chaparral, desert chaparral is more open. Many species of the semi-desert chaparral such as chamise, bigberry manzanita, holly leaf cherry (*Prunus ilicifolia*), and shrub interior live oak (*Quercus wislizeni*) are also common in more coastal xeric chaparral—except the desert variant has unique shrub species such as Mojave ceanothus (*Ceanothus greggii* var. *vestitus*), curl leaf mountain mahogany, and flannelbush (*Fremontodendron californicum*). Several chaparral shrub species, such as flannelbush, ashy silktassel (*Garrya flavescens*), cup leaf ceanothus, and chamise's shaggy cousin the redshank, occur on the drier desert-facing slopes and are neither members of the mesic nor xeric chaparral groups. These, along with such scrub oak species such as Muller oak (*Quercus cornelius-mulleri*) and pointleaf manzanita (*Arctostaphylos pungens*), are members of an interior desert-margin chaparral, which ranges further south and well inland, skipping across some of the higher desert mountains all the way to the rimrock country of central Arizona and south as far as the Sierra Madre of northwestern Mexico. This chaparral receives less winter rain and relatively more summer rain, although total rainfall is probably substantially less than chaparral on the Pacific Ocean side of the mountains. Interspersed between the woody, tall shrubs are shorter sub-shrubs of rabbitbrush (*Ericameria* spp.), hairy yerba santa (*Eriodictyon trichocalyx*), and California buckwheat (*Eriogonum fasciculatum*), along with desert species such as valley cholla (*Cylindropuntia californica* var. *parkeri*), pancake prickly pear (*Opuntia chlorotica*), and low clonal forms of Joshua tree.

Conservation of the Landscape

Many vegetation types in the South Coast Ranges face an uncertain future. Continued population growth in southern California, climate change, modifications to historic fire regimes, and the spread of invasive plant and animal species threaten the existence of some types and guarantee others will change in composition.

Between 1990 and 2010 the population of southern California grew from 14.6 to 16.5 million. This growth has been accompanied by a sharp increase in human-caused fires and unnaturally short periods between fires in many vegetation types. Chaparral, for example, is resilient to burn intervals of between 20 and 150 years, but intervals of less than 15 years are now common. These truncated intervals threaten the persistence of obligate seeders whether they are shrubs (many *Ceanothus* species) or trees (cypresses and serotinous pines) (Syphard et al. 2009).

Much urbanization has occurred at the contact zone where human development intermingles with intact vegetation. Development at this wildland-urban interface has not only increased ignitions but also has encouraged the spread of exotic grasses into recently burned areas. Once exotics become established in the early post-fire years, they increase the likelihood that another fire will re-burn the area and eliminate re-establishing native shrub seedlings. This positive feedback between fire and exotics incrementally eliminates the native vegetation and converts it to annual grasslands that can tolerate frequent fires. A similar exotic grass–interval fire scenario has impacted desert vegetation throughout the West. The spread of the exotic annual cheatgrass (*Bromus tectorum*) has created an understory fuel bed that readily carries fire at short intervals. Many desert shrub and tree species are poorly adapted to frequent fire and, like many chaparral shrubs, are disappearing (Chambers et al. 2007).

Climate change poses yet another threat to the native vegetation. For example, computer simulations indicate the climatically suitable habitat of Engelmann oak in San Diego County will gradually shrink and eventually disappear perhaps causing the extinction of this endemic oak (Conlisk et al. 2012). Mixed conifer forest vegetation in the San Jacinto Mountains is moving upward in elevation and subalpine and alpine vegetation will gradually decrease in area and perhaps be eliminated altogether. And, as southern California becomes warmer and drier, the threat of large wildfires may increase in the San Bernardino Mountains.

Mark Borchert
Deveree A. Kopp

Top: The wildland-urban interface has increased the frequency of wildfires that threaten the persistence of chaparral.
Todd Keeler-Wolf

Above: Recurring fires in short time intervals and invasive annual plants threaten the chaparral landscape of the San Jacinto Wildlife Area, Potrero Unit, in Riverside County.
Julie M. Evens

CHAPTER THIRTEEN

Mojave Desert

The Mojave Desert lies primarily within southeastern California, although southern Nevada, western Arizona, and even a tiny portion of southwestern Utah are a part of this desert ecosystem (EPA 2007). Over 17% of California is considered part of the Mojave Desert, putting it on par with the Sierra Nevada as a major portion of the California landscape. In the classic sense, true deserts are extremely dry—less than 100 mm of precipitation per year—and have less than 3% plant cover. California deserts are wetter than this and they average more plant cover, so they are technically "semi-deserts" (MacMahon 1979) and can be distinguished by predominantly shrubby life forms with varying density.

While the Mojave is the smallest of three warm North American deserts, it has a significant amount of geographical and topographical complexity, affording greater ranges in temperature, precipitation, and elevation than any other North American desert. Witness Death Valley, the lowest point on the continent, and the adjacent Inyo and Panamint mountains, among the highest found in deserts of North America. The western tip of the Mojave, in Antelope Valley, has a Mediterranean climate similar to eastern Los Angeles, with just slightly less annual rainfall and slightly lower mean winter temperatures. Conversely, Death Valley

Above: Representative of the Mojave Desert are island-like mountains interrupting massive bajadas of interfingering alluvial fans with semi-desert vegetation, including creosote bush, brittle bush, and white bursage.
Julie M. Evens

Opposite: From mountain range to basin, Death Valley National Park provides extremes in topography and complexity of vegetation, here looking from the Panamint Range with montane shrublands and singleleaf pinyon woodlands into Death Valley with the Badwater Basin salt flats and alkali sink vegetation.
Dana York

Top: Urbanization of the Antelope Valley and other parts of the Mojave Desert has fragmented several types of once common regional vegetation including Joshua tree and California juniper woodlands.
Todd Keeler-Wolf

Above: Old lake beds, or playas, like Bristol Lake occur as vast expanses draining separate basins of the Mojave Desert.
Julie M. Evens

*Left: Amargosa niterwort (*Nitrophila mohavensis*), is an endangered species, endemic to moist alkaline flats of lower Carson Slough and Ash Meadows.*
Jim André

has the lowest precipitation in North America and ranks among the highest summer temperatures anywhere on the planet. From west to east, across the Mojave, precipitation varies significantly from a strongly winter-dominant pattern to a bimodal pattern with near-equal rainfall in summer and winter months.

The Mojave Desert is also adjacent to the most heavily populated region of the state. Lancaster, Palmdale, Pearblossom, Hesparia, Victorville, and several other rapidly growing cities are home to commuters who work in Los Angeles' urban center. They have expanded explosively into the surrounding desert landscape in the past 30–40 years. Just east of the California–Nevada border lie the lights and sprawl of Las Vegas. Vast open spaces of the Mojave Desert are managed and protected by the Bureau of Land Management (BLM) as wilderness areas, or as national park lands within Death Valley National Park, Joshua Tree National Park, and Mojave National Preserve. However, the western and southern regions of the Mojave Desert have had more human-mediated impacts than the mountainous and more remote eastern portions. Many of the Mojave's valleys are being tapped for water, are planned for solar energy development, or have already become part of an agricultural and suburban patchwork. The most heavily compromised portion, the western Mojave, is home to some of the greatest annual wildflower displays in western North America.

Mojave vegetation is complex, varied, and anything but monotonous. It has a high number of endemic species as compared to most ecoregions in the U.S. (Rickets et al. 1999). These endemics include ancient lineages and newly evolving ones. The Mojave's diversity of vegetation is related to its major landform features.

Imagine you are flying low above the

Mojave. You pass over broad fans of accumulated sediments forming aprons of gently sloping coalescing alluvial fans (or bajadas) around the base of rugged mountains. These fans spread outward as alluvium-filled valleys and into closed basins separating range after range of mountains. Shining white and perfectly flat, dry lakes (or playas) sit at the bottom of most of the closed basins. The mountains bordering these basins are made of granite, or ancient layered and striped metamorphic rocks, calcium-rich limestone, and marble, and are peppered with recent dark lava flows and cinder cones. The highest mountains are in the northeast and were uplifted as part of the fault-block ranges of the border between the Mojave and the Great Basin regions, contrasting with the lower relief of the western Mojave.

Ecologically, a gradient of stressful conditions extends from the middle of saline basins, up through fine, sandy, lower alluvial fans and dunes, then across upper alluvial fans and bajadas with coarser soils, up onto the rocky lower mountain slopes, and finally up to the highest peaks, where the vegetation may be locally dominated by pines and junipers.

The most restricted and sometimes simplest desert vegetation patterns occur on the salty playas and basins, where severe high and low temperatures, high salinities, and extremely variable flooding limit the species variety. The most widespread desert vegetation occurs on the alluvial fans or bajadas, which experience less intense stresses. The higher mountains, although only covering less than 20% of the landmass of the Mojave (Norris and Webb 1990), are home to the vast majority of different vegetation types. For example, over 80% of all the alliances in the central Mojave are largely restricted to the mountains (Thomas et al. 2004; Keeler-Wolf 2007).

Immense topographic relief is typical of the northern Mojave. Soda Lake in deep Saline Valley sits at below 600 m while Bristlecone pine grow atop the nearby Inyo Mountains looming at over 3,350 m only a few miles away.
TODD KEELER-WOLF

BASINS, PLAYAS, SINKS, AND ALKALINE MARSHES

Closed basins in the Mojave Desert contain fine-grained, saline-alkaline soils where over 200 individual dry lakebeds occur. A playa is usually salt-encrusted and devoid of plants, and the water table below is saline or brackish. Toward the edge of the playa, intermittent patches of halophytes (salt-tolerant plants) grow on soil mounds raised a few centimeters or decimeters above playa level. Further away from the playa center, these patches increase in size and density, and are gradually replaced by saltbush (*Atriplex* spp.) scrub as the playa touches the lower parts of the surrounding bajadas. Upward on the gentle toe of the surrounding slopes, the saltbush scrub is gradually replaced

Top: *The most common vegetation in the Mojave Desert is creosote bush and white bursage scrub, as seen along Death Valley Road.*
Robert Wick

Above: *Iodine bush regularly occurs in basins of Death Valley with high salinity and available water tables.*
Julie M. Evens

by creosote bush (*Larrea tridentata*) scrub or other communities.

Alkali sinks are among the harshest habitats due to high salinity, variable water availability, and high summer temperatures. Two main shrub species, iodine bush (*Allenrolfea occidentalis*) and bush seepweed (*Suaeda moquinii*), characterize alkali sink vegetation throughout much of the Mojave. Both usually occupy fine-textured saline soils, though iodine bush is the more salt-tolerant of the two.

Arrow weed (*Pluchea sericea*) associated with iodine bush, bush seepweed, salt grass (*Distichlis spicata*), and other halophytes occurs in two distinct settings in the Mojave. Arrow weed may form dense thickets around permanent alkali seeps, or it may be more open in the large intermittent sinks, such as the floor of Death Valley or Saline Valley. The famous Devil's Cornfield in Death Valley is named after evenly spaced shrubs of arrow weed that have formed tufted pedestals as a result of

sand accreting at their bases.

The water table may be close enough to the surface to support marshes dominated by American bulrush (*Schoenoplectus americanus*), common reed (*Phragmites australis*), Cooper's rush (*Juncus cooperi*), yerba mansa (*Anemopsis californica*), alkali sacaton (*Sporobolus airoides*), and salt grass. These marshes are scattered around the floor and edges of desert basins, and may be associated with faults—which conduct water to the surface of the desert.

Mesquite Woodlands

Both honey mesquite (*Prosopis glandulosa*) and screwbean mesquite (*P. pubescens*) form scattered stands as far north as Saline Valley. Mesquite stands are widespread, adjacent to drainages, seeps, and playas with subterranean water; they also occur on sand dunes and dune sheets where plants can tap water within the sand. Screwbean mesquite stands appear to require surface water

Top: Arrow weed thickets in the Saline Valley.

Above: Interior desert basins with high water tables also support emergent marsh and alkaline wetland vegetation including salt grass and alkali sacaton.
Julie M. Evens photos

Top: Honey mesquite forms coppices in the dunes, often interspersed with fourwing saltbush.
Todd Keeler-Wolf

*Above: Broom baccharis (*Baccharis sergiloides*) occurs around springs and rocky arroyos in the lower hills and mountains.*
Julie M. Evens

of lower salinity (Turner et al. 1995). This species often mixes with willows (*Salix exigua, S. gooddingii*) or Fremont cottonwood (*Populus fremontii*), and is sometimes part of riparian desert vegetation, as along the Amargosa River east of Death Valley.

Since mesquite stands rely on a permanent subterranean water supply, their health is a good indicator of shifts in ground water level. Several large stands have been reduced and largely eliminated over the past several decades by ground water pumping, including those at Kohn (western Mojave) and Mesquite (central Mojave) dry lakes.

Riparian Vegetation

Freshwater riparian vegetation is scattered about streams, rivers, and in canyons where soils are permanently saturated or temporarily flooded. Small areas of riparian vegetation are associated with some mountains in the northern Mojave, but the most extensive riparian stands occur along the Mojave River in the vicinity of Victorville and along the Colorado River north of Needles. Both areas historically included expansive stands of Fremont cottonwood and tall willows. Immediately adjacent to the active river channels, where water flows during most winter and spring seasons, shrubby sandbar willow (*Salix exigua*) and mulefat (*Baccharis salicifolia*) are common. Along river terraces are dense Fremont cottonwood forests with an understory turf of native creeping rye grass (*Leymus triticoides*). Farther away from active channels are open Fremont cottonwood stands with an understory of allscale (*Atriplex polycarpa*). In some parts of the flood plain, invasive exotic Russian olive (*Elaeagnus angustifolia*) has become very common.

Riparian vegetation north of Needles is now frequently dominated by non-native tamarisk (*Tamarix* spp.) thickets. Where reservoirs and slow moving water exist, native bulrushes (*Schoenoplectus acutus, S. americanus*), cattails (*Typha domingensis, T. latifolia*), and common reed form marshes.

Some of the most diverse riparian stands occur in the higher desert mountains, for example in the Panamint, Cottonwood, Argus, and Inyo mountains of the northern Mojave. Stands of willows (*Salix exigua, S. laevigata*) and several species of *Baccharis* (*B. emoryi, B. salicifolia, B. sergiloides*) grow along with desert olive (*Forestiera pubescens*), and Fremont cottonwood, with desert wild grape (*Vitis girdiana*) clambering over them. A few small springs in the high elevation eastern Mojave have remnant individuals of netleaf hackberry (*Celtis reticulata*), a small tree with eastern affinities.

Saltbush Scrub

Saltbush species occur in varied combinations mostly in basins and lower bajadas across the Mojave. Allscale has the broadest salt tolerance of all the local saltbushes, and consequently its range is the most extensive. It occurs in and around sandy washes on lower bajadas and around moderately saline playas. In contrast, shadscale (*Atriplex confertifolia*) occurs in both uplands and around playas, but is closely tied to

the cooler upper elevations. Desert holly (*A. hymenelytra*) has a largely upland distribution, characterized by very dry, often hot and sterile lower slopes. Fourwing saltbush (*A. canescens*), the least salt-tolerant of the group, grows in the deep sand of washes or dunes. Quailbush (*A. lentiformis*) is a large deep-rooted shrub, and is most abundant where groundwater is shallow and soil salinity is relatively high as along ditches, drainages, and playas.

Vegetation of Dunes and Sand Sheets

Plants that exist on shifting sand dunes have their own distinct morphological traits. Here grow deep tap-rooted annuals, stoloniferous grasses, and perennials and shrubs with the ability to layer, or send their roots into the water reserves deep within the dunes. Average total plant cover on dunes and sand sheets is low except in good rainfall years, when annual cover can be spectacular with wildflowers including desert sand verbena (*Abronia villosa*), evening primrose (*Oenothera deltoides*), desert needles (*Palafoxia arida*), desert twinbugs (*Dicoria canescens*), big galleta grass (*Pleuraphis rigida*), desert panic grass (*Panicum urvilleanum*), and many others.

Mojave dunes contain unique stands of Eureka Valley dune grass (*Swallenia alexandrae*) and Eureka Dunes evening primrose (*Oenothera californica* ssp. *eurekensis*), sandpaper plant (*Petalonyx thurberi*), and Thurber's beardstongue (*Penstemon thurberi*).

Top left: Desert olive forms dense verdant thickets around seeps and springs in the desert. Todd Keeler-Wolf

Top right: Allscale is the most widespread saltbush, ranging from rocky volcanic hills to broad intermontane valleys.

Above: Shadscale scrub on calcareous upland slopes with a kaleidoscope of wildflowers. Julie M. Evens photos

Top: Dune systems may be ablaze with blooms in a good rainfall year and include evening primrose, desert needles, desert twinbugs, other native annuals, and big galleta grass along Devils Playground in the Mojave National Preserve.
Julie M. Evens

Top right: Eureka Valley dune grass grows only on the Eureka Dunes at the northern end of Death Valley National Park.

Above: Brittle bush is common along with creosote bush on south-facing hillslopes in the central and southern Mojave.
Todd Keeler-Wolf photos

Vegetation of Bajadas, Hills, and Washes

Creosote bush scrub, typically along with the shorter and more compact white bursage (*Ambrosia dumosa*), covers about two thirds of the Mojave Desert, sprawling over millions of acres of bajadas, hills, and washes. The creosote bush zone ranges from monospecific, sparsely-vegetated scrub in rocky, dry and hot lower bajadas, to diverse stands with over 30 perennial species on the cooler and better-watered upper bajadas, at the interface with montane vegetation.

Although creosote bush is the single most characteristic shrub of all North American deserts, it in combination with white bursage is characteristic of the Mojave Desert and adjacent northwestern Sonoran Desert. Stretching from Antelope Valley in the west to Eureka Valley in the north, and ranging south into the Sonoran Desert, creosote bush and white bursage co-dominate vast areas of sandy alluvial fans and bajadas, as well as rocky uplands and slopes. Brittle bush (*Encelia farinosa*) often replaces white bursage on very hot and dry southerly facing slopes in the southern Mojave.

Desert holly thrives in the hottest and most arid portions of the Mojave.

Usually, it is the only conspicuous shrub present though often sparse in cover. It also forms stands along alkaline sink edges, on fine Pleistocene lakebed sediments, and on dark, coarse basaltic slopes in the Saline Range in the northern Mojave, where its white foliage contrasts starkly with the dark basalt rock.

Desert washes form complex patterns across the lower fans and bajadas. They are often lined with larger and more densely growing shrubs, which take advantage of the occasional torrents of water coursing down them during wet winters or brief summer cloud-bursts. Lower elevation washes of the southern Mojave have much in common with the adjacent Colorado Desert, containing smoke tree (*Psorothamnus spinosus*), desert lavender (*Hyptis emoryi*), and in a few locations, taller small-leaved trees such as ironwood (*Olneya tesota*) and blue palo verde (*Parkinsonia florida*). The upper elevation washes of the Mojave contain an array of species that merge

Top: The vast landscape of the Mojave is dominated by a combination of creosote bush and its shorter associate, white bursage, shown here with spring wildflowers in the Mojave National Preserve.
Todd Keeler-Wolf

Above: Desert holly is characteristic of the most harsh settings in the desert, including the dark volcanic cinders of Ubehebe Crater in Death Valley National Park.
Julie M. Evens

with the adjacent sagebrush (*Artemisia* spp.), singleleaf pinyon (*Pinus monophylla*, and juniper (*Juniperus* spp.) vegetation of the Great Basin Desert.

Probably the most widely distributed vegetation type of Mojave washes is cheesebush (*Hymenoclea salsola*) scrub. Cheesebush, an excellent colonizer, is often the dominant shrub of wash systems throughout low elevations of the desert. It has occupied parts of the Mojave for over 11,000 years (Cole 1986), is frost- and drought-tolerant, and has a short lifespan and high seed production (Webb et al. 1988). It may be equally at home in recently burned upper bajada scrub, or in heavily grazed or other degraded landscapes, because it can sprout after disturbance and has fruits adapted to both flood and wind dispersal.

Other widespread wash alliances in the central and southern Mojave include desert willow (*Chilopsis linearis*) woodland, catclaw acacia (*Acacia greggii*) thorn scrub, black-stem rabbitbrush (*Ericameria paniculata*) scrub, and Virgin River brittle bush (*Encelia* [*actonii, virginensis*]).

Top: Smoke tree is characteristic of lower elevation washes in the southern Mojave Desert. Here, the large shrubs are surrounded by Arizona lupine (Lupinus arizonicus).
JULIE M. EVENS

Above: Cheesebush is the most widespread shrub of washes and other disturbed areas.
JIM ANDRÉ

Above: Desert willow has showy tubular flowers which bloom in washes throughout much of the Mojave.
Aaron Schusteff

Left: Catclaw acacia, blooming with clusters of yellow flowers, occurs in washes across the Mojave.

Below: Black-stem rabbitbrush scrub occurs along sandy washes, flowering in the fall, at Mojave National Preserve.
Julie M. Evens photos

Ephemeral Herbaceous Vegetation

Herbaceous vegetation occurs throughout the Mojave on the lower bajadas and fans. It may be non-evident after years of poor rainfall, then burst forth in profusion in a good El Niño year. These "wildflower years" are spectacular events and draw many tourists to the desert.

Desert pavement is a surface layer of close-fitting small rocks, largely devoid of places where germinating plants can reach into the soil. In many cases these surfaces, varnished with dark minerals, have been in place for hundreds of thousands of years and have slowly built up an armor of rocks and solidified soil impenetrable to most plants. This difficult habitat nonetheless supports

Top: Lupines, poppies, phacelias, purplemat (Nama demissum), *and other wildflowers are common in El Niño years.*
Julie M. Evens

Above: Desert-gold produces spectacular displays in El Niño years on what otherwise are largely unvegetated desert pavements.
Todd Keeler-Wolf

extensive seed banks of showy desert-gold (*Geraea canescens*) or suncups (*Camissonia* spp.), which brighten thousands of acres of otherwise harsh inhospitable bajadas throughout the central and eastern desert in good wildflower years. Devil's spineflower (*Chorizanthe rigida*), several annual buckwheats (*Eriogonum* spp.), and a variety of *Phacelia* species show up regularly on these desert pavements.

In the more Mediterranean climate of the western Mojave, many annuals are shared with adjacent central and southern California, including California poppy (*Eschscholzia californica*), common goldfields (*Lasthenia gracilis*), lupine (*Lupinus* spp.), and fiddleneck (*Amsinckia* spp.). Annuals put on spectacular displays during adequate rainfall years in Antelope Valley and adjacent areas. As in cismontane California, many stands, once dominated by showy flowers in spring, may be dominated later in the summer by non-native annuals or, less commonly, perennial native grasses such as desert needlegrass (*Achnatherum speciosum*).

Upper Bajada Vegetation

Moving up from the lower bajadas and basins the extensive creosote bush–white bursage scrub gives way to a narrower zone characterized by Joshua tree (*Yucca brevifolia*) woodland, Mojave yucca (*Y. schidigera*) scrub, mixed Mojave scrub, cactus gardens, and also black brush (*Coleogyne ramosissima*) scrub. This is the most floristically diverse part of the Mojave.

The range of Joshua tree outlines the extent of the Mojave Desert in California, southern Nevada, and northwestern Arizona. This tree-sized lily typically makes up relatively low cover, but it is a significant indicator species of the upper elevation portion of Mojave Desert scrub.

Mojave yucca occupies similar niches, but appears to require more reliable summer moisture and is capable of

Above: Extensive California poppy stands are characteristic of the western Mojave, west of Lancaster.
Neal Kramer

Far left: Desert needlegrass forms extensive bunchgrass grasslands in only a very few places. One of the most extensive is in the western Antelope Valley.
Julie M. Evens

Left: Large colorful wasps known as tarantula hawks (Pepsis sp.) are effective pollinators of several milkweeds (Asclepias spp.) in the Mojave, such as desert milkweed (A. erosa) in the western Mojave, Kern County.
Todd Keeler-Wolf

tolerating slightly higher summer temperatures. Unlike the Joshua tree, it does not occur in the western or northern Mojave where summer rain is less significant. Mojave yucca and banana yucca (*Yucca baccata*) are also tolerant of calcareous soils and regularly occur on exposed outcrops of limestone in California's eastern Mojave, often with an array of other unusual endemic "calcicole" shrubs including Death Valley joint fir (*Ephedra funerea*), Utah mortonia (*Mortonia utahensis*), and Stansbury cliff rose (*Purshia stansburiana*).

Black brush-Desert Chaparral-California Juniper Zone

Black brush is an endemic shrub of the American southwest. Black brush scrub forms a distinct belt above the creosote bush zone and below the "cold desert" singleleaf pinyon and big sagebrush (*Artemisia tridentata*) zone, forming relatively dense and monospecific stands that show up dark gray against the lighter backdrop of the mountain soils. It tends to dominate on rocky or shallow soils on upper bajadas, pediments, and rocky slopes in the transition zone between the warm deserts (Mojave and Sonoran) and the cool Great Basin.

For a small non-sprouting shrub, black brush is extremely long-lived with ages in excess of 400 years reported (Webb et al. 1988). It is typically killed by fire—and recovery is slow since seed

Top: Joshua tree woodland quintessentially characterizes the mid elevations in the region.

Middle: Mojave yucca scrub commonly occurs at the mid elevations with a diversity of associated plants.
Todd Keeler-Wolf photos

Left: Disjunct stands of Utah mortonia occur along mid-elevation calcareous slopes in the Mojave Desert.
Julie M. Evens

dispersal from nearby stands is required for re-establishment. Sinuous, sharp transitions between remnant stands of black brush and adjacent stands that may have been burned over 50 years ago are common sights throughout most of its range in California. Unfortunately, black brush scrub overlaps with the core habitat of the recently established, invasive red brome (*Bromus rubens*). In summers after moderate to high rainfall years, red brome forms tinder-dry fuel that carries fire through black brush stands, causing significant reduction throughout its range.

Spiny hop sage (*Grayia spinosa*) and Anderson's boxthorn (*Lycium andersonii*) scrubs are considered intermediate in climate between the creosote bush–white bursage scrub of the lower Mojave Desert, and the black brush scrub in the higher elevation zone. Extensive stands of spiny hop sage scrub, Anderson's boxthorn, and the Nevada joint fir (*Ephedra nevadensis*) scrub occur in the Panamint and other desert mountains of Inyo County where the soils are moderately sandy to loamy and the climatic conditions appear similar to the black brush scrub zone.

Another distinctive vegetation type in this zone is the spiny menodora (*Menodora spinescens*) scrub, a low thorny shrub scattered throughout mountains of the northeastern Mojave and adjacent southern Great Basin Desert, on shallow, well-drained, gravelly and rocky slopes. Despite its occupation of the transitional upper Mojave Desert climatic belt, spiny menodora is apparently frost sensitive. Webb et al. (2003) report a severe reduction of spiny menodora cover, probably as a result of a hard freeze in 1990.

Along the western boundary of the Mojave, in the Transverse, Peninsular, and southern Sierra Nevada, relatively dry and open chaparral vegetation occurs at the upper edges. An assortment of evergreen shrubs mixes with drought-deciduous shrub species. This desert chaparral interfingers at its upper extent with mixtures of Great Basin woodland, which includes singleleaf pinyon and California or Utah juniper (*Juniperus californica, J. osteosperma*), discussed further in Chapter Five, Mono and Southeastern Great Basin.

Desert chaparral resembles the open stands of cismontane California

Top: Black brush darkens the upper slopes of the Granite Mountains just below the singleleaf pinyon belt.
Todd Keeler-Wolf

Above: Nevada joint fir scrub with Cooper's goldenbush occurs commonly at mid to upper elevations.
Julie M. Evens

Top: Spiny menodora scrub is characteristic of rocky mid-elevation desert slopes in the northern Mojave and southeastern Great Basin, here surrounding a small grove of Joshua trees.

Above: Singleleaf pinyon woodlands occur in mountains of the Mojave primarily above 1,200 m.
JULIE M. EVENS PHOTOS

chaparral, but is characterized by species endemic to this desert transition zone, many of which extend to southern Arizona and adjacent Mexico. Three species of scrub oaks are dominants in localized areas: Tucker oak (*Quercus john-tuckeri*) in western Mojave, Tehachapi Mountains, and inner south Coast Ranges; Muller oak (*Q. cornelius-mulleri*) in the eastern Transverse and Peninsular ranges; and Sonoran live oak (*Q. turbinella*) in the eastern Mojave, on into Utah and Arizona. Manzanitas are associated with these scrub oaks and include bigberry manzanita (*Arctostaphylos glauca*) in the southern and western Mojave to central Coast Ranges, and pointleaf manzanita (*A. pungens*) in the eastern Mojave to inner central Coast Ranges and northern Mexico. Other widespread evergreen shrubs include cup leaf ceanothus (*Ceanothus greggii*), ashy silktassel (*Garrya flavescens*), hollyleaf redberry (*Rhamnus ilicifolia*), and sugarbush (*Rhus ovata*).

Stands also may be dominated by California juniper, but contain a similar understory of drought-deciduous desert shrubs. Most of the junipers are large shrubs with a stature similar to the scrub oaks. However, they are often very evenly spaced, whereas the oaks and other desert chaparral sclerophylls appear clumped. Unfortunately, extensive stands of California juniper in Antelope Valley of the western Mojave have been reduced by fire and clearing for agriculture.

When fire, grazing, and other clearing disturbances take place in this zone, several naturally occurring, shorter-lived shrubs grow in the years immediately following until the slower-growing longer-lived shrubs return to the site. Many short-lived shrubs, such as rabbitbrush (*Ericameria* spp.), tend to quickly colonize from adjacent unburned lands. In the western Mojave and locally in other parts of the desert, narrow-

leaf goldenbush (*E. linearifolia*) or rubber rabbitbrush (*E. nauseosa*) are common dominants. Similar settings in the eastern Mojave are often dominated by Cooper's goldenbush (*E. cooperi*) or needleleaf rabbitbrush (*E. teretifolia*). In particular, needleleaf rabbitbrush scrub from both the Mojave and the western Sonoran desert transition belt in the Peninsular Ranges seems to be replacing some of the recently burned California juniper stands (Keeler-Wolf 2007).

A different transitional state following lightning-caused fires involves the temporary replacement of fire-killed black brush by desert needlegrass. Desert needlegrass may prevail for several decades following a fire until black brush and other shrubs gradually recolonize. Brooks and Matchett (2003) have noted this trend as a former natural pathway prior to the introduction and widespread dominance of red brome in this zone.

Vegetation Persistence and Conservation

The distant and sweeping vistas and harsh climatic conditions confer a timeless aspect to the Mojave Desert. Yet its current vegetation patterns arose fairly recently following radical climatic changes over just the past 10,000 years. Evidence supporting the relative youth of species in the Mojave Desert is provided by numerous studies of packrat (*Neotoma lepida*) middens—where sequential layers of vegetation have been accumulated by packrats and have been fossilized in the dry habitat, embedded in sterile ureic acid for thousands of years. Mesic juniper woodlands have been replaced by xeric juniper communities—and juniper has been replaced by mixed scrub and black brush scrub—within the last 7,500 years in the Great Basin–Mojavean transition area. Evidence from packrat middens also shows that the now-ubiquitous creosote bush invaded the Mojave Desert only about 9,000 years

*Above: California juniper stands occur as wooded scrublands, such as in the Antelope Valley with California buckwheat (*Eriogonum fasciculatum*).*
Todd Keeler-Wolf

Below: Disjunct stands of Sonoran live oak scrub are found in the New York Mountains and Midhills of the Mojave National Preserve where summer monsoonal storms are common.
Julie M. Evens

Top: Fire scars show clearly on the slope of Clark Mountain and persist for decades, much to the overall detriment of the black brush scrub.

Top right: Desert tortoise.
TODD KEELER-WOLF PHOTOS

Above: Fields of mirrors and solar power towers from Ivanpah Solar Electric Generating System are evident from the edge of Mojave National Preserve in Ivanpah Valley.
JULIE M. EVENS

ago, and singleleaf pinyon and juniper woodlands have moved upslope hundreds of meters from previously occupied basins during the same period. Change toward a more arid climate has forced some expansions in plant ranges, and even new species or subspecies have appeared. Local desert features, if present at all during the close of the last glacial period, were confined to the driest sites such as on the eastern "rain shadow" of high ranges and on hot, sunny south-facing slopes (Raven and Axelrod 1995).

So, despite the long evolutionary history of some individual species of shrubs, the current plant groupings have come about relatively recently and are certainly continuing today. The broad extent of certain species combinations that we now see speaks to the combined abilities of colonizers and ecological generalists. With continuing climatic shifts, other more narrowly ranging species are likely to become restricted to smaller and smaller refugia and eventually go locally extinct. Equilibrium and stability in many of these communities has yet to be reached. In fact, substantial shifts in species composition of stands are known to have taken place less than 50 years ago due to the broadly fluctuating climate (Webb et al. 2003).

Such realizations should increase our awareness of the need for long-range conservation strategies of Mojave vegetation such as sustaining broad intact habitats for specialized desert plants and animals that live within them, including Mojave ground squirrel (*Xerospermophilus mohavensis*) and desert tortoise (*Gopherus agassizii*).

Opposite top: Gradient of vegetation from the base of the Tehachapi Mountains runs from creosote bush to Joshua tree, California juniper, and singleleaf pinyon woodlands, to canyon live oak and white fir atop distant mountains.

Opposite right: The ecological gradient between cismontane, montane, and desert has become fragmented in recent years by large wind farms with their network of access roads, fences, and rotating blades lethal to raptors and bats.
TODD KEELER-WOLF PHOTOS

While large-scale regional planning projects have been initiated throughout the Mojave, many of these plans have not had sufficient local focus on the varied mosaics of vegetation. They also are conflicted by demands for a multiplicity of uses, including grazing, off-highway motor vehicle use, private land development, and solar and wind energy development. The combined effects of unnatural fire and invasive plants, the excessive grazing of much of the desert, and other stresses, such as off-highway vehicle use, nitrogen deposition, and atmospheric carbon dioxide increases, will continue to change the composition and dynamics of Mojave vegetation. Appropriate management responses, such as maintaining natural corridors between largely intact habitats of the ecoregion, must become an established practice.

Julie M. Evens
Todd Keeler-Wolf

CHAPTER FOURTEEN

Sonoran-Colorado Desert

The Colorado Desert, the southernmost of California's deserts, is named for its association with the historic Colorado River Delta. It is a part of the warm Sonoran Desert that also covers much of southern Arizona, Baja California, and the state of Sonora, Mexico. Physiographically, the area is composed of the Salton Trough to the west, and to the east a series of broad alluvial fans, or bajadas, that drain eastward toward the lower Colorado River Valley. The Colorado River forms the eastern boundary of California's Colorado Desert. To the northeast the Colorado Desert grades into the southern Mojave Desert, primarily along an elevational gradient of approximately 350–500 m, with the two deserts divided by the San Bernardino, Little San Bernardino, and Eagle Mountains. To the east, as the elevation drops toward the Colorado River, the Colorado Desert extends northward along the Colorado River Valley, reaching the vicinity of Needles. Its western boundary abuts the steep Peninsular Ranges. From the Los Angeles Basin, Interstate 10 descends eastward to the desert through San Gorgonio Pass between the formidable San Bernardino and San Jacinto Ranges.

Geography, Geology, and Climate

The Peninsular Ranges create a rain shadow that intercepts winter storms from the Pacific Ocean. As a result the Colorado Desert differs from much of the Mojave Desert by receiving a significant portion of its annual precipitation

Above: The Colorado River at the eastern boundary of California's Colorado Desert, here south of Needles, is bordered with honey mesquite, southern cattail, American bulrush (Schoenoplectus americanus), *and tamarisk patches.*
Jim André

Opposite: The Santa Rosa and San Jacinto Mountains National Monument, rising precipitously from the desert floor, supports California fan palm oases and other quintessential desert flora.
Robert Wick

Top: The Algodones Dunes are California's largest dune system, shown here in the Buttermilk Valley with creosote bush in the background.
Gina Radieve

Above: The San Jacinto Mountains, northernmost of the Peninsular Ranges, form a rain shadow that intercepts precipitation from the west. Still, with adequate spring or summer rains, seasonal wildflowers including desert dandelion fill in the gaps of creosote bush scrub on this alluvial fan.
Allan Schoenherr

from summer monsoonal storms, with total precipitation ranging from about 50 to 300 mm per year depending on location and elevation. These monsoonal storms are infrequent, but can be intense and localized. The proportion of summer precipitation increases toward the east, reaching 35% of the annual mean along the Colorado River (Shreve 1925), but may reach 65% of the total at any given location (Tevis 1958). On the western border of the Colorado Desert, convectional storms from the south produce summer rain that increases in volume with elevation and falls primarily on the eastern slopes of the Peninsular Ranges. Records from Deep Canyon in the Santa Rosa Mountains indicate approximately one third of precipitation occurs in August and September, ranging from 130 mm at Indio to 350 mm per year at higher elevations (Zabriskie 1979). Summer temperatures frequently reach 45° C, and the mean daily maximum may exceed 38° C for several weeks (Shreve and Wiggins 1964).

The Salton Trough, a low elevation rift valley associated with the San Andreas Fault, is the largest below-sea-level basin in the Western Hemisphere (Oglesby 2005). Repeatedly filled and emptied by the Sea of Cortez and freshwater lakes, the Salton Sea is California's largest inland body of water and currently lies in the lowest point of the valley. The Salton Sea is the second lowest point in the United States, second to Death Valley only by 2 m. Its surface level hovers around 68 m below sea level and its sea floor is 84 m below sea level. The Salton Trough north of the Salton Sea is known as the Coachella Valley; to the south it is called the Imperial Valley. Irrigated by water from the Colorado River carried to the area by the All-American Canal, these valleys have been converted into

some of the richest agricultural lands in California. The southeastern portion of the Imperial Valley contains California's largest continuous dune field, the Algodones Dunes. At 64 km long and 10 km wide these dunes extend southward all the way to Baja California.

Today the source water for the Salton Sea is primarily irrigation runoff. The Salton Basin is separated from the Gulf of California by alluvial deposits of the Colorado River which are only about 10–12 m higher than the Sea of Cortez. By meandering back and forth over time, the mouth of the river alternately has drained into the basin and filled it with water, or as it does today, drained into the Sea of Cortez. On the western side of the Salton Trough, at the base of the Santa Rosa Mountains, about 60 m above the present level of the Salton Sea, is a clearly distinguishable old shoreline of former freshwater Lake Cahuilla that dried up about 500 years ago. Painted Canyon in the Mecca Hills east of Indio, and Painted Gorge north of Interstate 8, contain a series of colorful lakebeds that also bear witness to repeated inundations of the Salton Trough. Similarly, scenic lakebeds along Salt Creek Canyon on the south side of the Orocopia Mountains contain such a record.

Anza-Borrego Desert State Park, west of the Salton Sea, lies in a region of high biological diversity characterized by old lakebeds (badlands), sand dunes, bajadas, and rocky slopes. Its many habitats include well-established marsh and riparian vegetation associated with San Felipe Creek, one of the few permanently running, natural watercourses in the Colorado Desert. When precipitation is adequate, spring episodes are marked by bands of blooming shrubs and spectacular displays of wildflowers. Ranging from

Top: Red Canyon, a branch of Salt Creek Canyon in the Orocopia Mountains, with a smoke tree wash woodland at the base.

Above: The waterline of Lake Cahuilla, which dried up about 500 years ago, is still apparent.

ALLAN SCHOENHERR PHOTOS

Top: The Yuha Desert has been designated by the Bureau of Land Management as an Area of Critical Environmental Concern. Robert Wick

Above: The badlands of Anza-Borrego Desert State Park. Allan Schoenherr

the Vallecito Badlands eastward, fossils in sedimentary beds have yielded a remarkable record of plant and animal fossils that depict a nearly continuous record of climate change over the last seven million years (Jefferson and Lindsay 2005).

South of Anza-Borrego and south of Interstate 8 lies a region of rolling hills and washes that extend southward into Baja California. This area, known as the Yuha Desert, has been designated by the Bureau of Land Management as an Area of Critical Environmental Concern (ACEC) because of archeological significance and sensitive species. East of the Salton Trough, running from north to south, lie the Little San Bernardino, Cottonwood, Orocopia, and Chocolate Mountains. Where a prevailing wind and a source of sand occur, dune fields may form. These areas are located in the Coachella Valley near Palm Desert, in Anza-Borrego Desert State Park, along Palen Lake east of Joshua Tree National

Park, and in the Algodones Dunes south of Glamis. These are areas where spring wildflowers put on a spectacular visual display in those years when precipitation and temperature conditions are favorable, and they are home to scores of endemic species and specialized plants. The Algodones Dunes in particular are home to the largest number of dune-endemic plants in North America.

Top: Sand dunes along Palen Lake support specialized plants and animals, such as evening primrose, Palmer's wislizenia (Wislizenia refracta *ssp.* palmeri) *and Mojave fringe-toed lizard* (Uma scoparia).
Julie M. Evens

Above: Creosote bush scrub on an alluvial fan below the Orocopia Mountains.
Allan Schoenherr

Vegetation

By virtue of its southerly location and low elevation, overall, the Colorado Desert is the warmest and driest of California's three desert regions. Unlike the well-known landscape of southern Arizona, which may be dominated by saguaro (*Carnegiea gigantea*) and other tall cacti, the Colorado Desert is visually dominated by sparsely vegetated low mountains and ancient alluvial fans dotted by creosote bush (*Larrea tridentata*). As elevation increases to

Top: *Wildflower display with desert sand verbena and desert-gold in the Coachella Valley.* Allan Schoenherr

Above: *Allscale scrub growing near the edge of Palen Lake.* Julie M. Evens

the west, creosote bushes become larger and on the upper alluvial fans, tree species such as blue palo verde (*Parkinsonia florida*) and ironwood (*Olneya tesota*) form a sparse woodland. The rocky slopes along the western and northern borders of the desert are dominated by succulent species such as cacti and agaves (*Agave* spp.) intermixed with drought-deciduous plants such as brittle bush (*Encelia farinosa*) and ocotillo (*Fouquieria splendens*). While precipitation and temperature are the primary environmental features that regulate the distribution and patterns of desert plants, the chemical and structural composition of the soil has a profound influence as well. Along a slope or bajada in a typical desert basin there is a predictable gradient of soils from very coarse and rocky on the upper alluvial fans, to fine and saline in the basin centers. On hillsides above the sediments are rocky slopes that represent erosional surfaces from which sediments are derived. Thus, in a gradient from the salty clay lakebed soils to the upper hillsides, water percolation rates become increasingly more rapid so that water availability is of shorter duration farther up the slopes. The strategies and adaptations of desert plants are associated with their position on these slopes.

Playas

Increased salinity and reduced soil aeration limit the distribution of creosote bush in the vicinity of playas and salt flats. These areas are dominated by halophytes, salt-tolerant plants, most of which are in the goosefoot family (Chenopodiaceae). In the drier saline soils of the Colorado Desert, dominant alliances include fourwing saltbush (*Atriplex canescens*) scrub and allscale (*A. polycarpa*) scrub. On wetter, highly saline soils, typical saltbushes are replaced with succulent-leaved shrubs. This habitat, also known as alkali sink, is found locally along the Colorado River, Salton Sea, Palen Lake, and Clark Lake in Anza-Borrego Desert State Park. In the wettest areas of saline soils, iodine bush (*Allenrolfea occidentalis*) scrub occurs, and where salinities range from 0.5–1.0%, bush seepweed (*Suaeda moquinii*) scrub occurs. This habitat is discussed in more detail in Chapter 13, Mojave Desert.

Alluvial Fans

Halophytes occur in a ring around the playas. Above this zone, on the coarsely textured and porous alluvial fans and lower hills, the dominant vegetation is typically composed of drought-enduring, long-lived perennial shrubs that retain leaves on a year-round basis. The abundance of this habitat accounts for the astounding abundance of creosote bush scrub in California's warm deserts. The most widespread vegetation type in the region, and in California, is the creosote bush–white bursage (*Larrea tridentata–Ambrosia dumosa*) scrub. The U.S. Bureau of Land Management's (2002) report on the northern and eastern Colorado Desert indicated that this community covers 69% of the study area. This vegetation dominates vast intermountain bajadas above the saltbush scrub and below the rocky slopes and hillsides. These shrubs dominate on well-drained soils with low total salinity (Marks 1950)

Abundance of creosote bush–white bursage scrub may be a function of age. There is evidence that creosote bush inhibits germination of white bursage

Top: Iodine bush scrub scattered along the edge of the salt-encrusted Palen Lake.

Above: Creosote bush scrub is very common on alluvial fans and lower hills, and larger blue palo verde and ironwood are dotted along the intersecting washes on the fans.

Julie M. Evens photos

Top: The creosote bush–white bursage scrub alliance is the most widespread vegetation type in the region.
Julie M. Evens

Above: African mustard invasion in Anza-Borrego Desert State Park.
Allan Schoenherr

seeds. It appears that single-species stands of creosote bush scrub where bursage is conspicuously absent, have been undisturbed for long periods of time. It has been demonstrated that white bursage, which has a shallower root system and more rapid growth, is more likely to colonize disturbed areas. Through time, creosote bush becomes established, utilizes ground water more efficiently, and outlives white bursage, effectively replacing it as a co-dominant species.

Other plant species appear with creosote bush scrub where environmental conditions are appropriate. For example, the root parasites known as rhatanies (*Krameria grayi* and *K. erecta*) are fairly common in upper or undulating bajadas. Also, where sand accumulates between

shrubs, certain plants such as longleaf joint fir (*Ephedra trifurca*), crinklemats (*Tiquilia* spp.) and big galleta grass (*Pleuraphis rigida*) become established. During years with appropriate precipitation, the gaps between the creosote bushes fill with annual wildflowers, particularly smooth desert dandelion (*Malacothrix glabrata*). Unfortunately these gaps also may fill with non-native annual grasses (*Bromus rubens, Schismus* spp.) and in recent years, with African mustard (*Brassica tournefortii*). This invasion is particularly apparent in Anza-Borrego Desert State Park. When these ephemeral plants die they may become fuel that enables fire to be carried from shrub to shrub, ultimately changing the character of creosote bush scrub which is not truly fire-adapted.

Rocky Uplands

On sunny slopes of upper bajadas and rocky hillsides, where soil is coarse and water is available for shorter periods of time, a mixture of drought-deciduous and succulent species often grows. The most diverse and distinctive vegetation type of the Colorado Desert is located here (Schoenherr 1992, Schoenherr and Burk 2007).

Many of the drought-deciduous plants are leafless during the dry season, a form of drought dormancy or evasion. However, they provide a palate of color to desert slopes when in full abundance and bloom during the spring following a wet winter. Ocotillo is a conspicuous plant that occurs throughout the Sonoran Desert, and may be considered an indicator species for the region. Ocotillos produce tubular red

Above: Drought-deciduous and succulent plants of the cactus scrub community tend to dominate on rocky hillsides.
Allan Schoenherr

Below: Anza-Borrego Desert State Park is an important refuge for the endangered peninsular population of desert bighorn sheep (Ovis canadensis nelsoni).
Robert P. Sikora

Above: Ocotillo, considered to be an indicator species for the Sonoran Desert region, flowering at Coyote Canyon, Anza-Borrego Desert State Park.

Below: A cactus wren (Campylorhynchus brunneicapillus) *nest in a cholla.*
ALLAN SCHOENHERR PHOTOS

Opposite top: Brittle bush on rocky uplands in the Turtle Mountains.
JIM ANDRÉ

Opposite bottom: Barrel cactus and teddy bear cholla in rocky habitat of the Chuckwalla Mountains.
ALLAN SCHOENHERR

flowers pollinated by hummingbirds, and on years with adequate precipitation, a red band of blooming ocotillos my cover entire hillsides. The ability of ocotillo to produce new leaves in quick response to precipitation is legendary. Five to six episodes of leaf gain and loss in a single year have been reported (Shreve and Wiggins 1964). New leaves emerge at the base of the spines within 24–48 hours of watering (Killinbeck 1990). Succulent-like, ocotillos store water in their stems and carry on photosynthesis with chlorophyll in the bark during leafless periods.

Brittle bush scrub is a common alliance on the rocky uplands. This silver-leaved shrub in the sunflower family produces a cloak of bright yellow flowers in the spring. Typical of many sunflower species, these blossoms track the sun as it moves across the sky. From a southerly perspective, a slope dominated by brittle bush will be a blaze of yellow color, as all the blossoms face you. Other common drought-deciduous species include bladdersage (*Salazaria mexicana*), bricklebush (*Brickellia desertorum*), and indigo bush (*Psorothamnus schottii*). Of limited distribution, but just as showy in flower, are several species of sage. Santa Rosa sage (*Salvia eremostachya*) occurs in the Santa Rosa Mountains, the Orocopia sage (*S. greatae*) is limited to the Orocopia and Chocolate Mountains, and wand sage (*S. vaseyi*) occurs on the desert sides of the Peninsular Ranges.

Succulent species such as agaves and cacti absorb water quickly and store it in their tissues for use during the rest of the dry season. Cold winter temperatures, which commonly drop below freezing in the Mojave, are an important limiting factor for northward distribution of succulent species. Thus, the warm

Sonoran Desert is characterized by more species of cactus than the Mojave. About 125 cactus species occur in the Sonoran Desert proper, and at least 22 are in the Colorado Desert (Ingram 2008).

Cacti bloom following the winter rainy period when the plants are fully hydrated. Cactus flowers are unusually large and showy, relying on the spectacle of color, rather than odor, to attract pollinators. Individual blossoms for the most part, however, are open for only a day or so (Nobel 2002). Colorado Desert species have large open flowers, which are usually bee-pollinated. Other types of cacti, particularly those with white flowers, may be moth- or bat-pollinated.

The largest common cacti of the Colorado Desert are the barrel cactus (*Ferocactus cylindraceus*) and the chollas such as teddy bear (*Cylindropuntia bigelovii*), Munz's (*C. munzii*), and Wolf's

Top: Teddy bear cholla scrub with brittle bush and other shrubs at Yaqui Wells, Anza-Borrego Desert State Park.
Julie M. Evens

Above: Common fish-hook cactus. Note the fish-hook spines and disproportionately large flowers for this small cactus.
Allan Schoenherr

(*C. wolfii*). The barrel cactus may reach 3 m in height, but smaller specimens are more typical. These cacti seem perfectly suited to grow in unusual or unsuspecting places, such as in the cracks of boulders. The barrel cactus is further notable by being able to tolerate the highest temperature of any California cactus (Ingram 2008). Another barrel type of cactus is known as many-headed barrel cactus (*Echinocactus polycephalus*). One large specimen in the Little Maria Mountains northwest of Blythe was composed of approximately 150 heads.

There are at least nine species of cholla in the Colorado Desert, although teddy bear cholla patches are the most ubiquitous alliance. It is an indicator for the Colorado Desert, and it often occurs as dense, nearly pure stands. Stands may be the result of sheet flooding that carries the stem joints into areas that have been denuded of vegetation by the flood. Teddy bear cholla is the famous "jumping cholla" which, aided by barbed spines, inadvertently snags itself on someone's shoe. This species rarely reproduces sexually with flowers, but reproduces asexually by breaking apart into stem joints that lie in great abundance on the ground.

The Colorado Desert has four species of small cacti that resemble pincushions. The fish-hook cacti are distinguished by having a number of hooked spines. Growing close to the ground, fish-hook cacti are subject to intense summer heat. Three species occur in the Colorado Desert; California fish-hook cactus (*Mammillaria dioica*); common fish-hook cactus (*M. tetrancistra*); and Graham's fish-hook cactus (*M. grahamii*), an Arizona species that only occurs in the Whipple Mountains in California.

Desert agave (*Agave deserti*) scrub is

commonly found on upper bajadas and rocky hillsides along the western side of the Colorado Desert. There are upland and wash terrace stands, the latter being more concentrated due to fluvial disturbance which initiates asexual clones. Another subspecies (*A. deserti* ssp. *simplex*) occurs in the northeastern part of the desert, from the Whipple Mountains to Needles. It also occurs in the Mojave National Preserve, but nowhere are the stands as extensive as the western Colorado Desert in Anza-Borrego Desert State Park. Desert agave is sometimes called "century plant" because it lives for many years before it blooms, though on average it might take 15–25 years before it blooms. South of the international border its common pollinator is the lesser long-nosed bat (*Leptonycteris yerbabuenae*). Interestingly, in the population of desert agaves near the Mexican border, the reproductive parts of the flowers are shorter, facilitating pollination by the carpenter bees (*Xylocopinae* spp.) (Fulton 1984). Desert agaves bloom once and die, as if the reproductive effort exhausts the parent plant. They are very common, however, as they reproduce asexually by growing a series of offspring or pups around the base of the parent plant. Many parts of desert agave are valued as food by desert animals and humans. Young flower stalks and flowers are eaten raw and the entire rosette of succulent leaves may be roasted and eaten.

Jojoba (*Simmondsia chinensis*), an evergreen shrub found in the rocky uplands, has relatively succulent leaves. Its most distinguishing feature is its wax-filled

Top: Desert agave scrub in bloom.
JORDAN ZYLSTRA

Right: Lesser long-nosed bats feeding on agave nectar.
MERLIN TUTTLE, MERLIN TUTTLE'S BAT CONSERVATION

Top: Smoke tree woodland in a wash east of the Eagle Mountains.

Above: Bird's nest in a graythorn shrub.

Julie M. Evens photos

fruit and attempts have been made to farm the shrubs for the high-quality wax, which for the most part have been met with misfortune.

Desert Wash and Microphyll Vegetation

Another feature that distinguishes the Colorado Desert is the abundance of trees, particularly along washes. These trees are important components of the Sonoran Desert in general, although most do not occur in other California deserts. This habitat, sometimes called microphyll woodland, contains several trees in the legume family (Fabaceae). Some of these, such as blue palo verde and smoke tree (*Psorothamnus spinosus*), are drought-deciduous. Although smoke trees do occur in the southern Mojave, their distribution is said to outline the Colorado Desert. Ironwood keeps leaves all year long, and the mesquites (*Prosopis* spp.) are winter-deciduous. These deep-rooted species are able to reach down to tap soil moisture most of the year. A threat to these habitats are the non-native tamarisks (*Tamarix* spp.), and public agencies responsible for maintenance of our public lands have undertaken programs to eliminate these invasive, water-demanding species. In Arizona, microphyll woodland often includes large columnar cacti such as the saguaro, which in California only occurs in the Whipple Mountains near the Colorado River.

Smoke tree woodlands typically are found in the most active portion of desert washes while blue palo verde–ironwood woodlands occur directly adjacent in less active washes and upper bajadas that intermittently receive sheet flows. Smoke tree and palo verde are leafless most of the time, but by virtue of chlorophyll in the bark and deep root systems are able to carry on 40–75% of their photosynthetic needs in the absence of leaves. Ironwood

retains leaves on a year-round basis, but loses them a few at a time during the dry season. These three species seem restricted mostly to the Colorado Desert because their northward dispersal is limited by cold winter temperatures. In contrast, desert willow (*Chilopsis linearis*) is a cold-tolerant member of the tropical catalpa family (Bignoniaceae), and its alliance is common in active desert washes of the Sonoran and Mojave Deserts.

Two common drought-deciduous shrubs also occur as alliances in these desert washes. Chuparosa (*Justicia californica*) has tubular red flowers that attract hummingbirds. Desert lavender (*Hyptis emoryi*), in the mint family (Lamiaceae), is an aromatic shrub with blue flowers. These shrubs attract swarms of bees, and other wildlife that utilize these shrublands, including nesting and foraging songbirds. Less common, but widely dispersed are the strikingly thorny shrubs of graythorn (*Ziziphus obtusifolia* var. *canescens*) and spiny abrojo (*Condalia globosa*). When in full leaf, their bitter juicy fruits are blackish to purplish.

Riparian Vegetation

The California fan palm (*Washingtonia filifera*) oasis, dominated by California's only native palm, grows in California and Arizona along fault systems of desert mountains. Subterranean water in desert canyons and washes comes near the surface, providing a permanent water supply. Coyotes (*Canis latrans*) and birds eat the fruit, and when they take refuge in the cool oases during warm weather they introduce the seeds through their feces. Thus the palms are carried from one oasis to another. Also, Native Americans are known to have eaten the fruit (dates) so it is entirely possible that the seeds were carried by them and planted in various oases in the southwest.

California fan palm oases are the only habitat in the Colorado Desert in which

Top: Blue palo verde woodland lines a desert wash on the Chukwalla Bench.

Above: Chuparosa, a drought-deciduous, hummingbird-pollinated shrub, is common along the edges of desert washes. Allan Schoenherr photos

fire is a natural factor. Lightning starts fires in palms, and may kill individual trees, but most of the fires clear out the underbrush without killing trees. By clearing out competition, fires increase surface water and remove the understory vegetation allowing palm seedlings to establish. It is possible that Native Americans deliberately set fires as well.

The Mexican fan palm (*W. robusta*) is not native to California, but has been planted extensively as a horticultural variety and will hybridize with the California native. It differs from the California fan palm by its more slender trunk and greater height. Over their full length, Mexican palm petioles are sharply toothed, and the distinctive thread-like fibers that characterize the leaves of California fan palm are greatly reduced. The blue palm (*Brahea armata*) does not occur naturally in California, but its native distribution in Baja California occurs within 25 km of the international border on the east side of the Peninsular Ranges (Rebman and Roberts 2012). The blue palm has a stout trunk and the attractive leaves are a grayish blue-green color.

The palms are associated with winter-deciduous riparian species such as Fremont cottonwood (*Populus fremontii*) and black willow (*Salix gooddingii*). A common shrub in this habitat is arrow weed (*Pluchea sericea*). Typical marsh plants such as southern cattail (*Typha domingensis*) and spiny rush (*Juncus acutus*) may also be present in permanently wet areas. The Fremont cottonwood forest alliance, and certain willow (*Salix* spp.) alliances occur in corridors along permanent water courses, such as

Top: California fan palm oasis at Thousand Springs Oasis.
ALLAN SCHOENHERR

Left: Fremont cottonwood woodland in a permanent watercourse in Joshua Tree National Park.
JULIE M. EVENS

the Colorado River or San Felipe Creek, or in oases along with the palms. These are not true desert habitats, as permanent water enables these species to avoid true desert conditions.

Notable among winter deciduous species that also appear in washes as alliances are honey mesquite (*Prosopis glandulosa*), screwbean mesquite (*P. pubescens*), and catclaw acacia (*Acacia greggii*), the latter characterized by dense spikes of fragrant yellow flowers. The large root systems of mesquites are important stabilizers for sand dunes. In the Coachella Dunes, stabilized mounds up to 4 m in height may be associated with individual specimens of mesquite. Typical of all these legumes is the production of bean pods which contain nutritious seeds and many animals feed on the seeds when they are green. These seeds, however, when mature, have very hard seed coats that need to be scratched or scarified before water will penetrate. This feature is facilitated by summer thunderstorms that drop significant precipitation in a short period of time. Water coursing down the washes carries the seeds along with gravel and rocks that crack the coats, ensuring germination in wet soil. In order to stay alive, at first the seedling roots grow very rapidly and deeply, effectively maintaining contact with moist soil. The roots of some of these matured trees are legendary—one record places roots of honey mesquite in a mineshaft at a depth of 80 m (Solbrig and Orions 1977). However, a rule of thumb among well-diggers is that mesquite grows where water is no deeper than 10 m.

Sand Dunes

The Colorado Desert contains several dune fields, including the Coachella Dunes north of the Salton Sea, dunes at the mouth of Coyote Canyon in Anza- Borrego Desert State Park, dunes along Palen Lake east of Joshua Tree National Park, and the extensive Algodones Dunes on the southeastern part of the Imperial Valley. Distinctive vegetation adapts to the moving substrate of a dune field. Some

Top: Honey mesquite stabilizing sand dunes west of Palen Lake.
Julie M. Evens

Above: Colorado Desert buckwheat with its stalk-like base has adapted to the moving substrate of the Algodones Dunes.
Neal Kramer

Above: Displays of winter annual wildflowers are often dominated by sand verbena and desert poppy, such as this view of Coyote Canyon in Anza-Borrego Desert State Park.
ALLAN SCHOENHERR

Below: Sand food, with its distinctive shape and clusters of flowers.
MICHAEL L. CHARTERS

authors call plants of this community "psammophytic," with adaptations that enable them to live in shifting sands. The Algodones Dunes have six unique taxa. The perennial Colorado Desert buckwheat (*Eriogonum deserticola*) with its sinuous, stalk-like base, is a conspicuous dominant in this habitat. Sand food (*Pholisma sonorae*) is a distinctive endemic plant to these dunes, being a root parasite with the buckwheat as its host.

The dunes also abound with ephemeral wild flowers when environmental conditions are appropriate. After germination, these plants go through vegetative growth, flowering, and seed-set in one season. Ephemeral plants that germinate after winter rains are called winter annuals; summer annuals germinate in association with monsoonal storms during late summer. They grow rapidly and complete a lifecycle in a shorter period of time than winter ephemerals. Examples of summer annuals, which mostly occur in the southeastern part of the Colorado Desert, include windmills (*Allionia incarnata*), spiderlings (*Boerhavia wrightii*), and devil's claw (*Proboscidea parviflora*) with its dramatic hooked fruits.

In the Colorado Desert, most of the remarkable flower displays are of winter annuals such as desert sand verbena (*Abronia villosa*), evening primrose (*Oenothera deltoides*), desert poppy (*Eschscholzia parishii*), and desert needles (*Palafoxia arida*). In certain parts of these sandy habitats common plant associations are grouped into alliances. For example, even though desert twinbugs (*Dicoria canescens*) is inconspicuous, it occurs commonly in association with desert sand verbena, hence the alliance is known as the desert dunes (*D. canescens–A. villosa*) sparsely vegetated alliance. In

addition, the perennial big galleta grass forms a conspicuous alliance on sandy aprons. While desert lilies (*Hesperocallis undulata*) may appear along with winter annuals, they return from dormant bulbs rather than seeds.

Desert Pavement

In contrast to a dune field, in desert pavement areas the sand has blown away, leaving an armor of pebbles. Extensive desert pavements, found on very flat areas between washes and arroyos, are mostly devoid of perennial vegetation. Only a limited number of ephemerals are able to survive the extreme temperatures among the blackened pebbles and stones covered with desert varnish. Yet during good rain years annual plants abound, including desert-gold (*Geraea canescens*), pebble pincushion (*Chaenactis carphoclinia*), desert fivespot (*Eremalche rotundifolia*), and spineflowers (*Chorizanthe* spp.). One notable small plant, devil's spineflower (*Chorizanthe rigida*), emerges among the stones and produces broad green leaves as well as spines. After it dies, the brownish spiny vestiges of the plant remain for a year or so, giving the appearance of a small brown cactus.

Conservation Significance

Several pronounced mountain ranges stand as ecological islands within the boundaries of the Colorado Desert. The tallest of these, along the western and northern edges of the desert, reach sufficient elevation to support singleleaf pinyon (*Pinus monophylla*) and California juniper (*Juniperus californica*) in their upper reaches. Along Interstate 10, east of Chiriaco Summit and south of the highway, the Orocopia and Chuckwalla Mountains dominate the horizon. To the north lie the Eagle, Coxcomb, Palen, and McCoy Mountains. Between the moun-

Above: Big galleta grass is an important forage plant for large herbivores.
Deborah L. Stout

Below: Desert lily in the Chuckwalla Valley north of Desert Center.
Julie M. Evens

Top: Desert pavement on a wind-blown alluvial fan at the base of the Coxcomb Mountains above the Chuckwalla Valley.
Julie M. Evens

*Middle: Flat-tailed horned lizard (*Phrynosoma mcallii*), a candidate for state species protections, is threatened by a variety of human disturbances.*
Gary Nafis

Above: Mojave fringe-toed lizard, a species of special concern, at Iron Age Dunes. Long scales on hind feet facilitate traction when running on sand.
Allan Schoenherr

tains, broad plains and bajadas drain into Danby, Palen, and Ford Dry Lakes. East of the Chocolate Mountains, a large flat area, the Chuckwalla Bench, drains eastward toward the Colorado River. It is designated as an Area of Critical Environmental Concern because of its rich flora of approximately 160 plant species, including nine cacti and rare and endangered plants and animals such as Munz's cholla and the desert tortoise (*Gopherus agassizii*).

On the west side of the Colorado River are low but scenic mountain ranges of a volcanic nature. Running from north to south, these include the Sacramento Mountains near Needles, the Chemehuevi, Whipple, Riverside, and Palo Verde Mountains near Blythe, and the Picacho Mountains with picturesque Picacho State Recreation Area in Imperial County. The Whipple Mountains near Parker are particularly notable because of the presence of plants such as saguaro cactus, Graham's fish-hook cactus, and little-leaf palo verde (*Parkinsonia microphylla*), species normally associated with the Arizona portion of the Sonoran Desert.

Overall, many of the species here are shared with the Mojave and Arizona's Sonoran Deserts; however, about 100 plant species occur nowhere in California outside of the Colorado Desert (Mathias 1978). Many of these occur in families that are typically southern in distribution. Some are relictual, remnants of a formerly widespread distribution, and occur in specific microhabitats, including California fan palm at springs and seeps, chuparosa, California snakebush (*Colubrina californica*), Hall's tetracoccus (*Tetracoccus hallii*) on rocky slopes and within washes, and ragged rock flower (*Crossosoma bigelovii*) along rocky canyon walls. Some have very localized distribution, such as crown-of-thorns (*Koeberlinia spinosa*) in the Chocolate Mountains along washes and canyon walls, crucifixion thorn (*Castela emoryi*) associated with dry lakebeds, and elephant tree (*Bursera microphylla*), mostly in Anza-Borrego Desert State Park along rocky alluvial fans, washes, and slopes.

Some species that are typically associated with the Mojave, such as Mojave yucca (*Yucca schidigera*) and joint firs (e.g., *Ephedra viridis* and *E. nevadensis*), occur along upper bajadas and rocky slopes on the western and northern borders of the Colorado Desert at higher locations where cold temperatures mimic higher latitudes. The Colorado Desert differs from the Mojave by having a significant community of woodland trees that inhabit its washes and upper bajadas. Also, particularly to the west on rocky uplands, there is a unique assemblage of succulent and drought deciduous plants found nowhere else in California. There are more species of cacti in the Colorado Desert than in California's other deserts. This rocky upland habitat is well represented in Anza-Borrego Desert State Park, as are notable desert washes and oases of California fan palms. In spite of its significant aridity, the Colorado Desert is a region of great diversity, which has been afforded preservation by Joshua Tree National Park and the California State Park System.

In addition to invasive species and

illegal OHV activity, current threats in California's Sonoran–Colorado desert region include solar and wind power projects, such as the 800-hectare Genesis solar facility—already plagued by flooding at Ford Dry Lake, the Desert Sunlight solar facility along the eastern boundary of Joshua Tree National Park, the proposed Palen solar project near Palen Dry Lake, the Ocotillo wind farm near Anza-Borrego State Park, the San Gorgonio Pass Wind Farm east of Whitewater, and several wind projects proposed on the Chuckwalla Bench and near Indian Pass. The Blythe Mesa Solar Power Plant, in its initial stages, will be primarily built on previously farmed land across 1,400 hectares.

Allan A. Schoenherr

Top: The Whipple Mountains, in the farthest southeast part of California, are notable for the only populations of saguaro cactus in the state.
Robert Wick

Above: Ocotillo Express Wind Energy Facility, Imperial County, just south of Anza-Borrego Desert State Park, is one of several controversial desert energy projects complicated by biodiversity and cultural history issues.
Kim Clark

General References & Literature Cited

Foreword and Introduction

Bailey, R. G. 1995. *Description of the ecoregions of the United States,* 2nd edition. Miscellaneous Publication 1391. USDA Forest Service, Washington, DC.

Faber, P. M. 1997. *California's wild gardens,* First edition. California Native Plant Society, Sacramento, CA. [Reprinted 2005 by University of California Press].

Hilton, J. 1933. *Lost horizon.* William Morrow & Co., New York, NY.

Sawyer, J. O., T. Keeler-Wolf, and J. M. Evens. 2009. *A manual of California vegetation,* 2nd edition. California Native Plant Society, Sacramento, CA.

Chapter One: Northern California Coast

Albert, M. E., C. D'Antonio and K. A. Shierenbeck. 1997. Hybridization and introgression in *Carpobrotus* spp. (Aizoaceae) in California. I. Morphological evidence. *American Journal of Botany* **84**:896–904.

Atwater, B.F., S. Musumi-Rokkaku, K. Satake, Y. Tsuji, K. Ueda, K. and D.K. Yamaguchi. 2005. The orphan tsunami of 1700–Japanese clues to a parent earthquake in North America. Professional Paper 1707. U.S. Geological Survey, Reston, VA.

Axelrod, D. I. 1988. Outline history of California vegetation. Pages 139–193 *in* M.G. Barbour and J. Major, editors. *Terrestrial vegetation of California,* 2nd edition. Special Publication 9. California Native Plant Society, Sacramento, CA.

Barbour, M. G. 2007. Closed-cone pine and cypress forests. Pages 296–312 *in* M. G. Barbour, T. Keeler-Wolf and A. A. Schoenherr, editors. *Terrestrial vegetation of California,* 3rd edition. U.C. Press, Berkeley, CA.

Baker, H. G. 1972. A fen on the northern California coast. *Madroño* **21**:405–416.

Barry, W. J. and E. I. Schlinger, editors. 1977. Inglenook fen: a study and plan. Unpublished report. State of California, Resources Agency, Department of Parks and Recreation. Sacramento, CA.

Canter, A. 2012. Noteworthy collection. *Madroño* **59**:230.

Cascadia Geosciences. 2013. Tectonic land level changes and their contribution to sea-level rise, Humboldt Bay region, Northern California 2013 Status Update. Unpublished report. Cascadia Geosciences, McKinleyville, CA.

Cooper, W. S. 1958. *Coastal sand dunes of Oregon and Washington.* Geological Society of America Memoir 72. Boulder, CO.

Cooper, W. S. 1967. *Coastal sand dunes of California.* Geological Society of America Memoir **104**:1–147.

Gilkerson, W. 2008. A spatial model of eelgrass (*Zostera marina*) habitat in Humboldt Bay, California. Master's Thesis, Humboldt State University, Arcata, CA.

Green, S. 1999. Structure and dynamics of a coastal dune forest at Humboldt Bay, California. Master's Thesis, Humboldt State University, Arcata, CA.

Grewell, B. J. 2008. Hemiparasites generate environmental heterogeneity and enhance species coexistence in salt marshes. *Ecological Applications* **18**:1297–1306.

Grewell, B. J., J. C. Callaway, and W. R. Ferren, Jr. 2007. Estuarine wetlands. Pages 124–154 *in* M.G. Barbour, T. Keeler-Wolf and A.A. Schoenherr, editors. *Terrestrial vegetation of California,* 3rd edition. U.C. Press, Berkeley, CA.

IPCC. 2013. Technical Summary. Pages 33–115 *In* T. F. Stocker, D. Qin, G. K. Plattner, M. Tignor, S. K. Allen, J. Boschung, A. Nauels, Y. Xia, V. Bex and P. M. Midgley, editors. *Climate Change 2013: The Physical Science Basis. Contributions of Working Group I to the Fifth Assessment Report of the Intergovernmental Panel on Climate Change.* Cambridge University Press, New York, NY.

Lafferty, K. D., R. F. Hechinger, J. C. Shaw, K. Whitney, and A. M. Kuris. 2006. Food webs and parasites in a salt marsh ecosystem. Pages 119–134 *in* S. K. Collinge and C. Ray, editors. *Disease ecology: community structure and pathogen dynamics.* Oxford University Press, Oxford, U.K.

Levin, L. A., C. Neira, and E. D. Grosholz. 2006. Invasive cordgrass modifies wetland trophic function. *Ecology* **8**:419–432

Lagarde, L. A. 2012. Invasive *Spartina densiflora* Brongn. reduces primary productivity in a northern California salt marsh. Master's Thesis, Humboldt State University, Arcata, CA.

Leppig, G. T. 2002. A phytogeographic study of northern California peatlands. Master's Thesis, Humboldt State University, Arcata, CA.

Mitsch, W. J. and J. G. Gosselink. 2007. *Wetlands.* 4th edition. John Wiley & Sons. Hoboken, N.J.

Mitchell, M. L. 2012. A comparison of invertebrate communities occupying *Spartina*-invaded and restored salt marshes in Humboldt Bay, California. Master's Thesis, Humboldt State University, Arcata, CA.

Molinari, N., C. D'Antonio, and G. Thomson. 2007. *Carpobrotus* as a case study of the complexities of species impacts. *Theoretical Ecology* **4**:139–162.

Moore, J. E., M. A. Colwell, R. L. Mathis, and J. M. Black. 2004. Staging of Pacific flyway brant in relation to eelgrass abundance and site isolation, with special consideration of Humboldt Bay, California. *Biological Conservation* **115**:475–486.

National Academy of Sciences. 2012. *Sea-level rise for the coasts of California, Oregon, and Washington: past, present, and future.* The National Academies Press, Washington, DC.

Patton, J. R. 2004. Late Holocene coseismic subsidence and coincident tsunamis, southern Cascadia subduction zone, Hookton Slough, Wigi (Humboldt Bay), California. Master's Thesis, Humboldt State University, Arcata, CA.

Pickart, A. 2006. Vegetation of diked herbaceous wetlands of Humboldt Bay National Wildlife Refuge: classification, description and ecology. Unpublished report. U.S Fish and Wildlife Service, Arcata, CA.

Pickart, A. J. 2013. Dune restoration over two decades at the Lanphere and Ma-le'l Dunes in northern California. Pages 159–172 *in* M. L. Martinez, J. B. Gallego-Fernandez and P. A. Hesp, editors. *Restoration of Coastal Dunes.* Springer, NY.

Pickart, A. J. 2012. *Spartina densiflora* invasion ecology and restoration of native marshes at Humboldt Bay National Wildlife Refuge. Unpublished report. U.S Fish and Wildlife Service, Arcata, CA.

Pickart, A. J. and M. G. Barbour. 2007. Beach and dune. Pages 155–179 *in* M. G. Barbour, T. Keeler-Wolf and A. A. Schoenherr, editors. *Terrestrial vegetation of California,* 3rd edition. U.C. Press, Berkeley, CA.

Pickart, A. J. and J. O. Sawyer. 1998. *Ecology and restoration of northern California coastal dunes.* California Native Plant Society. Sacramento, CA.

Pickart, A.J., L.M. Miller, and T.E. Duebendorfer. 1998. Yellow bush lupine invasion in northern California coastal dunes: I. Ecological impacts and manual restoration techniques. *Restoration Ecology* **6**:59–68.

Shaughnessy, F.J., C.L. McGary, A. J. Frimodig, C. Witte, and G. B. Roberts. 2004. Known and unknown aspects of bottom-up and top-down regulation of eelgrass in Humboldt Bay, California. Proceedings of the Symposium: Current Perspectives on the Physical and Biological Processes of Humboldt Bay. March 15, 2004. Eureka, CA.

Sholars, R. E. 1984. The pygmy forest of Mendocino. *Fremontia* **12**:3–8.

Sholars, R. E. 1982. *The pygmy forest, and associated plant communities of coastal Mendocino County, California.* Black Bear Press. Mendocino, CA.

Sholars, T. 1997. Draft regional management plan for the Mendocino Coast pygmy forest. Report to California State Department of Parks and Recreation, College of the Redwoods, The Nature Conservancy, University of California Natural Reserve System, and The Trust for Wildland Communities. Mendocino, CA.

Tietz, J. R. and M. D. Johnson 2007. Stopover ecology and habitat selection of juvenile Swainson's Thrushes during fall migration along the northern California coast. *The Condor* **109**:795–807.

Vilà, M. and C. M. D'Antonio. 1998. Hybrid vigor for clonal growth in *Carpobrotus* (Aizoaceae) in coastal California. *Ecological Applications* **8**:1196–1205

Wiedemann, A. M. and A. J. Pickart. 1996. The *Ammophila* problem on the northwest coast of North America. Landscape and Urban Planning **34**:287–299.

Chapter Two: North Coast Ranges and Interior Ranges

Hickman, J. C. 1993. *The Jepson manual: higher plants of California.* University of California Press, Berkeley, CA.

Raiche, R. 2009. The Cedars: Sonoma County's hidden treasure. *Fremontia* **37**: 3–15.

Sawyer, J. O. 2006. *Northwest California: a natural history.* University of California Press, Berkeley, CA.

Chapter Three: Klamath Mountains

Sawyer, J. O. 2006. *Northwest California: a natural history.* University of California Press, Berkeley, CA.

Chapter Four: Modoc Plateau and Northwestern Basin and Range

Bailey, R. G. 1995. *Description of the ecoregions of the United States.* 2nd edition. Miscellaneous Publication 1391. USDA Forest Service, Washington, DC.

California Department of Fish and Game (CDFG). 2006. *California wildlife: conservation challenges.* Wildlife Health Center, University of California, Davis, CA.

D'Antonio, C. M. and P. M. Vitousek. 1992. Biological invasions by exotic grasses, the grass/fire cycle, and global change. *Annual Review of Ecology and Systematics* **23**:63–87.

Di Orio, A. P., R. Callas, and R. J. Schaefer. 2005. Forty-eight year decline and fragmentation of aspen (*Populus tremuloides*) in the South Warner Mountains of California. *Forest Ecology and Management* **206**:307–313.

Griffin, J. R. and W. B. Critchfield. 1972. The Distribution of Forest Trees in California. Research Paper PSW-82/1972. USDA Forest Service, Pacific Southwest Forest and Range Experimental Station, Berkeley, CA.

Hall, F. A. 1995. Determining changes in abundance of Sage Grouse *Centrocercus urophasianus* in California. California Department of Fish and Game, Wildlife Management Branch, Sacramento, CA.

Henstrom, M. A., M. J. Wisdom, W. J. Hann, M. M. Rowland, B. C. Wales, and R. A. Gravenmier. 2002. Sagebrush steppe vegetation dynamics and restoration potential in the interior Columbia Basin, U.S.A. *Conservation Biology* **16**:1242–1255.

Holland, R. F. 1998. No net loss? Changes in Great Valley vernal pool distribution from 1989 to 1997. California Department of Fish and Game, Natural Heritage Division, Sacramento, CA.

Holland, R. F. 2006. Mapping vernal pools and fens in the Modoc National Forest. Prepared for Modoc National Forest, Alturas, CA.

Keeler-Wolf, T., D. R. Elam, K. Lewis, and S. A. Flint. 1998. California vernal pool assessment preliminary report. California Department of Fish and Game, Sacramento, CA.

Miller, R. F. and J. A. Rose. 1999. Fire history and western juniper encroachment in sagebrush steppe. *Journal of Range Management* **52**:550–559.

Miller, R. F., T. J. Svejcar, and J. A. Rose. 2000. Impacts of western juniper on plant community composition and structure. *Journal of Range Management* **53**:574–585.

Miller, R. F. and R. J Tausch. 2001. The role of fire in pinyon and juniper woodlands: A descriptive analysis. Pages 15–30 in K. E. M. Galley and T. P. Wilson, editors. *Proceedings of the invasive species workshop: The role of fire in the control and spread of invasive species.* Misc. Pub. 11. Tall Timbers Research Station, Tallahassee, FL.

Miller, R. F., J. D. Bates, T. J. Svejcar, F. B. Pierson, and L. E. Eddleman. 2005. Biology, Ecology, and Management of Western Juniper (*Juniperus occidentalis*). Technical Bulletin 152, Oregon State University, Corvallis, OR.

Norris, R. M. and R. W. Webb. 1976. *Geology of California*, John Wiley & Sons, New York, NY.

Norton, J. B., T. A Monaco, J. M. Norton, D. A. Johnson, and T. A. Jones. 2004. Cheatgrass invasion alters soil morphology and organic matter dynamics in big sagebrush-steppe rangelands. Pages 57–63 in A. L. Hild, N. L. Shaw, S. E. Meyer, E. W. Schupp, and D. T. Booth, editors. *Seed and Soil Dynamics in Shrubland Ecosystems: Proceedings; 12th Wildland Shrub Symposium, August 12–16, 2002*, Laramie, WY. RMRS-P-31. Ogden, UT.

Novak, S. J. and R. N. Mack, 2001. Tracing plant introduction and spread: genetic evidence from *Bromus tectorum* (cheatgrass). *BioScience* **51**: 114–122.

Natural Resource Conservation Service (NRCS). 1993. *Soil Survey of Modoc National Forest Area, California*. Davis, CA.

Riegel, G. M., D. A. Thornburgh, and J. O. Sawyer. 1990. Forest habitat types of the South Warner Mountains, Modoc County, California. *Madroño* **37**: 88–112.

Riegel, G. M., R. F. Miller, C. N. Skinner, and S. E. Smith. 2006. Northeastern Plateaus Bioregion. Pages 255–263 in N. G. Sugihara, J. van Wagtendonk, K. E. Shaffer, J. Fites-Kaufman, and A. E. Thode, editors. *Fire in California's ecosystems*. University of California Press, Berkeley, CA.

Schaefer, R. J., D. J. Thayer, and T. S. Burton. 2003. Forty-one years of vegetation change on permanent transects in northeastern California: Implications for wildlife. *California Fish and Game* **89**:55–71.

Shultz, L. M. 2009. Monograph of *Artemisia Subgenus Tridentatae* (Asteraceae–Anthemideae). *Systematic Botany Monographs* **89**:1–131.

Smith, S. and B. Davidson. 2003. User's manual: terrestrial ecological unit inventory (TEUI) land type associations Modoc National Forest. Technical Publication R5-TP-015. USDA Forest Service, Pacific Southwest Region, Alturas, CA.

Smith, S. 1994. Ecological guide to eastside pine plant associations, northeastern California: Modoc, Lassen, Klamath, Shasta-Trinity, Plumas and Tahoe National Forests. Technical Publication R5-ECOL-TP-004. USDA Forest Service, Pacific Southwest Region, Alturas, CA.

_____. 1998. Riparian community type classification of national forest in northeastern California: first approximation. Unpublished report. USDA Forest Service, Pacific Southwest Research Station, Berkeley, CA.

The Nature Conservancy (TNC). 2001. Great Basin: an ecoregion-based conservation blueprint. Reno, NV.

U.S. Forest Service. 1991. Modoc National Forest land and resource management plan: record of decision and final environmental impact statement. USDA Forest Service, Pacific Southwest Region, Modoc National Forest, Alturas, CA.

_____. 2001. Sierra Nevada forest plan amendment: record of decision and final impact statement. Volumes 1–6. USDA Forest Service, Pacific Southwest Region, Vallejo, CA.

_____. 2010. Vegetation mapping data for the Pacific Southwest Region (GIS data layer). Remote Sensing Lab, McClellan, CA. Updated February 2010. Available at: http://www.fs.fed.us/r5/rsl/clearinghouse/gis-download.shtml. Accessed May, 2012.

U.S. Forest Service and Bureau of Land Management. 2012. Conservation strategy for *Orcuttia tenuis* on federal lands of the Southern Cascades and Modoc Plateau. Susanville, CA.

West, N. E. 1983. Western intermountain sagebrush steppe. Pages 351–374 in N. W. West, editor. *Ecosystems of the world. Volume 5: Temperate deserts and semi-deserts.* Elsevier Scientific Publishing Company, New York, NY.

Western Regional Climate Center (WRCC). 2012. Regional temperature and precipitation data. Available at: http://www.wrcc.dri.edu/. Accessed May, 2012.

Young, J. 2000. *Bromus tectorum*. In *Invasive plants of California's wildlands*. C. C. Bossard, J. M. Randall, and M. C. Hoshovsky, editors. University of California Press, Berkeley, CA.

Young, J. A., C. D. Clements, and H. C. Jansen. 2007. Sagebrush Steppe. Pages 587–608 in M. G. Barbour, T. Keeler-Wolf, and A. A. Schoenherr, editors. *Terrestrial vegetation of California*, 3rd edition. University of California Press, Berkeley, CA.

Chapter Five: Mono and Southeastern Great Basin (no References included)

Chapter Six: Sierra Nevada and Southern Cascades

Barbour, M. G., N. H. Berg, T. G. F. Kittel, and M. E. Kunz. 1991. Snowpack and the distribution of a major vegetation ecotone in the Sierra Nevada of California. *Journal of Biogeography* **18**:141–149.

Barbour, M. G. and W. D. Billings. 2000. *North American terrestrial vegetation*, 2nd edition. Cambridge University Press, New York, NY.

Barbour, M. G, T. Keeler-Wolf, and A. A. Schoenherr. 2007. *Terrestrial vegetation of California*, 3rd edition. University of California Press, Berkeley, CA.

Barbour, M. G., B. Pavlik, F. Drysdale, and S. Lindstrom. 1993. *California's changing landscapes*. California Native Plant Society, Sacramento, CA.

Barbour, M. G. and R. A. Woodward. 1985. The Shasta red fir forest of California. *Canadian Journal of Forestry* **15**:570–576.

Fites, J. 1993. Ecological guide to mixed conifer plant associations of the northern Sierra Nevada and southern Cascades. Technical Publication R5-ECOL-TP-001. USDA Forest Service, Pacific Southwest Research Station, San Francisco, CA.

Haller, J. R. and N. J. Vivrette. 2011. Ponderosa pine revisited. *Aliso* **29**:53–57.

Hunter, J. C. and M. G. Barbour. 2001. Through-growth of Douglas-fir; a model of rapid forest change without canopy gaps. *Journal of Vegetation Science* **12**:445–452.

Longstreth, C. 2014. Managing burned landscapes in the Sierra Nevada: back to the future (slowly). *Fremontia* **42**:7–13.

Parker, A.J. 1994. Latitudinal gradients of coniferous tree species, vegetation, and climate in the Sierran-Cascade axis of northern California. *Vegetatio* **115**:145–155.

Potter, D. A. 1998. Forested communities of the upper montane in the central and southern Sierra Nevada. General Technical Report PSW-GTR-169. USDA Forest Service, Pacific Southwest Research Station, Albany, CA.

Sawyer, J. O., T. Keeler-Wolf, and J. M. Evens. 2009, *A manual of California vegetation*, 2nd edition. California Native Plant Society, Sacramento, CA.

Sierra Nevada Ecosystem Project (SNEP). 1996. *Sierra Nevada Ecosystem Project: final report to Congress*, Volume 2. Assessments and scientific basis for management options. Water Resources Center Report No. 37. University of California, Centers for Water and Wildlands Resources, Davis, CA.

Weixelman, D. A., D. C. Zamudio, and K. A. Zamudio. 1999. Eastern Sierra riparian field guide. Technical Paper R4-Ecol.-99-01. USDA Forest Service, Intermountain Region, Ogden, UT.

Chapter Seven: Sierra Nevada Foothills

Alt, D. and D. W. Hyndman. 2000. *Roadside geology of Northern and Central California*. Mountain Press Publishing Company, Missoula, MT.

Fauver, T. 1998. *Wildflower walks and roads of the Sierra gold country*. Comstock Bonanza Press, Grass Valley, CA.

Hartwell, G. 2006. Ione plants. *Fremontia* **34**:19–24.

Mallek, C. 2009. Fire history, stand origins, and the persistence of Mcnab cypress, northern California, USA. *Fire Ecology* **5**:100–119.

Sawyer, J. O., D. A. Thornburgh, and J. R. Griffin. 1977. Mixed evergreen forest. Pages 359–382 *in* M. G. Barbour and J. Major, editors. *Terrestrial vegetation of California*. Wiley-Interscience, reprinted 1988 by the California Native Plant Society, Sacramento, CA.

Chapter Eight: Great Valley

Barbour, M. G., A. I. Solomeshch, and J. J. Buck. 2007. *Classification, ecological characterization, and presence of listed plant taxa of vernal pool associations in California*. Report to U. S. Fish and Wildlife Service, Agreement Study No. 814205G238, University of California, Davis, CA.

Barbour, M. G., A. Solomeshch, C. Witham, R. Holland, R. MacDonald, S. Cilliers, J. A. Molina, J. Buck, and J. Hillman. 2003. Vernal pool vegetation of California: variation within pools. *Madroño* **50**:129–146.

Barry, W. J. 1972. *The Central Valley prairie*. California Department of Parks and Recreation, Sacramento, CA.

Bartolome, J. W., W. J. Barry, T. Griggs, and P. Hopkinson. 2007. Valley Grassland. Pages 367–393 *in* Barbour, M. G., T. Keeler-Wolf, and A. A. Schoenherr, editors, *Terrestrial vegetation of California*, 3rd edition. University of California Press, Berkeley, CA.

Brewer, W. H. 1966. *Up and down California in 1860–1864*, 3rd edition. University of California Press, Berkeley, CA.

Buck-Diaz, J., J. Ratchford, and J. Evens. 2013. *California rangeland monitoring and mapping: focusing upon Great Valley and Carrizo Plain grassland habitats.* California Native Plant Society, Sacramento, CA.

Burtt Davy, J. 1898. Natural vegetation of alkali lands. *California Agricultural Experiment Station Report* **1895-97**:63–75.

Evett, R. R. and J. W. Bartolome. 2013. Phytolith evidence for the extent and nature of prehistoric Californian grasslands. *The Holocene* **23**:1644–1649.

Garone, P. 2011. *The fall and rise of the wetlands of California's Great Central Valley.* University of California Press, Berkeley, CA.

Germano, D. J., G. B. Rathbun, L. R. Saslaw, B. L. Cypher, E. A. Cypher, and L. M. Vredenburgh. 2011. The San Joaquin Desert of California: ecologically misunderstood and overlooked. *Natural Areas Journal* **31**:138–147.

Holland, R. F. 1978. *The geographic and edaphic distribution of vernal pools in the Great Central Valley, California.* Special Publication No. 3. California Native Plant Society, Berkeley, CA.

Holland, R. F. 1986. *Preliminary descriptions of the terrestrial natural communities of California.* California Department of Fish and Game, Sacramento, CA.

Holstein, G. 2001. Pre-Agricultural Grassland in Central California. *Madroño* **48**:254–264.

Holstein, G. 2011. Grasslands and prairies: what's in a name? *Fremontia* **39**:2-5.

Jepson, W. L. 1925. *Manual of the flowering plants of California.* Associated Students Store, University of California, Berkeley, CA.

Katibah, E. F. 1984. A brief history of riparian forests in the Central Valley of California. Pages 23–29 *in* R. E. Warner and K. M. Hendrix, editors. *California riparian systems: ecology, conservation, and productive management.* University of California Press, Berkeley, CA.

Keeler-Wolf, T., D. R. Elam, K. Lewis, and S. A. Flint. 1998. *California vernal pool assessment preliminary report.* The Resources Agency, California Department of Fish and Game. Sacramento, CA.

Keeler-Wolf, T., J. M. Evens, A. I. Solomeshch, V. L. Holland, and M. G. Barbour. 2007. Community classification and nomenclature. Pages 21–34 *in* M. R. Stromberg, J. D. Corbin, and C. M. D'Antonio, editors. *California grasslands: ecology and management.* University of California Press, Berkeley, CA.

McPhee, J. 1993. *Assembling California.* The Noonday Press, New York, NY.

Minnich, R. A. 2008. *California's fading wildflowers: lost legacy and biological invasions.* University of California Press, Berkeley, CA.

Preston, R. E. 2010. Alkaline rain pools: remnants of a vanishing landscape. *Fremontia* **37/38**:18–23.

Sawyer, J. O. and T. Keeler-Wolf. 1995. *A manual of California vegetation.* California Native Plant Society, Sacramento, CA.

Sawyer, J. O., T. Keeler-Wolf, and J. M. Evens. 2009. *A manual of California vegetation*, 2nd edition. California Native Plant Society, Sacramento, CA.

Schiffman, P. M. 2007. Species composition at the time of first European settlement. Pages 52–56 *in* M. R. Stromberg, J. D. Corbin, and C. M. D'Antonio, editors. *California grasslands: ecology and management.* University of California Press, Berkeley, CA.

Silveira, J. G. 2000. Vernal pools and relict duneland at Arena Plains. *Fremontia* **27/28**: 38–47.

Solomeshch, A. and M. Barbour. 2006. Defining restoration targets for the Great Valley Grassland State Park (San Joaquin Valley). *Grasslands* **1**:12–17.

Vaghti, M. G., and S. E. Greco. 2007. Riparian vegetation of the Great Valley. Pages 425–455 *in* M. G. Barbour, T. Keeler-Wolf, and A. A. Schoenherr, editors. *Terrestrial vegetation of California*, 3rd edition. University of California Press, Berkeley, CA.

Wester, L. 1981. Composition of the native grasslands in the San Joaquin Valley, California. *Madroño* **28**:231–241.

Chapter Nine: Central California Coast Ranges

Alexander, E. B., R. G. Coleman, T. Keeler-Wolf, and S. P. Harrison. 2006. *Serpentine geoecology of western North America: geology, soils, and vegetation.* Oxford University Press, New York, NY.

Baldocchi, D. D. and L. Xu. 2007. What limits evaporation from Mediterranean oak woodlands –The supply of moisture in the soil, physiological control by plants or the demand by the atmosphere? *Advances in Water Resources* **30**:2113–2122.

Borchert, M. 1985. Serotiny and cone-habit variation in populations of *Pinus coulteri* (*Pinaceae*) in the southern coastal ranges of California. *Madroño* **32**:29–48.

Borchert, M. I., N. D. Cunha, P. C. Krosse, and M. L. Lawrence. 1993. Blue oak plant communities of southern San Luis Obispo and northern Santa Barbara counties, California. General Technical Report PSW-139. USDA Forest Service, Pacific Southwest Research Station, Albany, CA.

Evens, J. M., A. Klein, J. Taylor, D. Hickson and T. Keeler-Wolf. 2006. *Vegetation classification and descriptions of the Clear Creek Management Area, Joaquin Ridge, Monocline Ridge, and environs in San Benito and western Fresno counties, California.* Report to USDI, Bureau of Land Management, Hollister District, California. California Native Plant Society and California Department of Fish and Game, Sacramento, CA.

Grossinger, R. M., C. J. Striplen, R. A. Askevold, E. Brewster, and E. E. Beller. 2007. Historical landscape ecology of an urbanized California valley: wetlands and woodlands in the Santa Clara Valley. *Landscape Ecology* **22**, Issue 1 Supplement: 103–120.

Ledig, F. T. 2000. Founder effects and the genetic structure of Coulter pine. *Journal of Heredity* **91**:307–315.

Plumb, T. R. and A. P. Gomez. 1983. Five southern California oaks: identification and postfire management. General Technical Report PSW-071. USDA Forest Service, Pacific Southwest Forest and Range Experimental Station, Berkeley, CA.

Prugh, L. R. and J. B. Brashares. 2012. Partitioning the effects of an ecosystem engineer: kangaroo rats control community structure via multiple pathways. *Journal of Animal Ecology* **81**:667–678.

Chapter Ten: Central California Coast

Brady, K. U., A. R. Kruckeberg, and H. D. Bradshaw. 2005. Evolutionary ecology of plant adaptation to serpentine soils. *Annual Review of Ecology, Evolution, and Systematics* **36**:243–266.

Cylinder, P. D. 1995. The Monterey ecological staircase and subtypes of Monterey pine forest. *Fremontia* **23**:7–13.

Ford, L. D. and G. F. Hayes. 2007. Northern Coastal Scrub and Coastal Prairie. Pages 180–207 *in* M. G. Barbour, T. Keeler-Wolf, and A. A. Schoenherr, editors. *Terrestrial vegetation of California*, 3rd edition. University of California Press, Berkeley, CA.

Hopkins, N. A. 1987. Mycorrhizae in a California serpentine grassland community. *Canadian Journal of Botany* **65**:484–487.

Kruckeberg, A. R. 1984. *California serpentines: flora, vegetation, geology, soils, and management problems.* University of California Press, Berkeley, CA.

Matthews, M. A. 2006. *An illustrated field key to the flowering plants of Monterey County and ferns, fern allies, and conifers.* Version 1.1. California Native Plant Society, Sacramento, CA.

Morgan, R. 1980. Locally unique native plants in the Santa Cruz County Coastal Zone. Plant list prepared for inclusion in the Santa Cruz County Local Coastal Program. Harvey and Stanley Associates, Inc., Alviso, CA.

Stromberg, M. R., P. Kephart and V. Yadon. 2001. Composition, invasibility, and diversity in coastal California grasslands. *Madroño* **48**:236–252.

VanDyke, E., K. D. Holl, and J. R. Griffin. 2001. Maritime chaparral community transition in the absence of fire. *Madroño* **48**:221–229.

Chapter Eleven: Southern California Coast

Axelrod, D. 1978. The origin of coastal sage vegetation, Alta and Baja California. *American Journal of Botany* **65**:1117–1131. *(Note: dates the last glacial retreat ca. 12,000 YBP)*

Brown, D. E. 1982. *Biotic communities of the American Southwest—United States and Mexico.* The University of Arizona, Tucson, AZ.

Dahl, T. T. 1990. Wetlands losses in the United States, 1780s to the 1980s. U. S. Fish and Wildlife Service, Washington, DC.

Grewel, B. J., J. C. Callaway, and W. R. Ferren, Jr. . 2007. Estuarine wetlands. Pages 124–154 *in* Barbour, M. G., T. Keeler-Wolf, and A. A. Schoenherr, editors. *Terrestrial vegetation of California*, 3rd edition. University of California Press, Berkeley, CA.

Ingram, S. 2008. *Cacti, agaves, and yuccas of California and Nevada.* Cachuma Press, Los Olivos, CA.

Jacobs, D., E. S. Stein, and T. Longcore. 2010. Classification of California estuaries based on natural closure patterns: templates for restoration and management. Technical Report 619. Southern California Water Research Project, Costa Mesa, CA. [Revised August 2011]

Junak, S. 2008. *A flora of San Nicolas Island California.* Santa Barbara Botanic Garden, Santa Barbara, CA.

Junak, S., D. A. Knapp, J. R. Haller, R. Philbrick, A. Schoenherr, and T. Keeler-Wolf. 2007. The California Channel Islands. Pages 229–252 *in* M. G. Barbour, T. Keeler-Wolf, and A. A. Schoenherr, editors. *Terrestrial vegetation of California*, 3rd edition. University of California Press, Berkeley, CA.

Mattoni, R. 1990. The endangered El Segundo blue butterfly. *Journal of Research on the Lepidoptera* **29**:229–304.

Minnich, R. A. 2008. *California's fading wildflowers: lost legacy and biological invasions.* University of California Press, Berkeley, CA.

O'Leary, J. F. 1989. Californian coastal sage scrub: general and considerations for Biological conservation. Pages 24–41 *in* A. A. Schoenherr, editor. *Endangered plant communities of southern California. Proceedings of the 15th Annual Symposium*, Special Publication No. 3. Southern California Botanists, Fullerton, CA.

Pickart, A. and M. Barbour. 2007. Beach and Dune. Pages 155–179 *in* M. G. Barbour, T. Keeler-Wolf, and A. A. Schoenherr, editors. *Terrestrial vegetation of California*, 3rd edition. University of California Press, Berkeley, CA.

Rundel, P. W. 2007. Sage scrub. Pages 208–228 *in* M. G. Barbour, T. Keeler-Wolf, and A. A. Schoenherr, editors. *Terrestrial vegetation of California*, 3rd edition. University of California Press, Berkeley, CA.

Sawyer, J. O., T. Keeler-Wolf, and J. M. Evens. 2009. *A manual of California vegetation, 2nd edition*. California Native Plant Society, Sacramento, CA.

Schoenherr, A. A. 1992. *A natural history of California*. University of California Press, Berkeley, CA.

Schoenherr, A.A. 2011. *Wild and beautiful: a natural history of open spaces in Orange County.* Laguna Wilderness Press, Laguna Beach, CA.

Schoenherr, A. A., C. R. Feldmeth, and M. J. Emerson. 1999. *Natural history of the islands of California*. University of California Press, Berkeley, CA.

Shreve, F. 1951. *Vegetation of the Sonoran Desert*. Publication 591. Carnegie Institute of Washington, Washington, DC.

Westman, W. E. 1981. Diversity relations and succession in Californian coastal sage scrub. *Ecology* **62**:170–184.

Westman, W. E. 1982. Coastal sage succession. Pages 91–98 *in* C. E. Conrad and W. C. Oechel, editors. Proceedings of the symposium on environmental consequences of fire and fuel management in Mediterranean ecosystems. General Technical Report PSW-58. USDA Forest Service, Pacific Southwest Forest and Range Experimental Station, Berkeley, CA.

Westman, W. E. 1983. Island biogeography: studies on the xeric shrublands of the inner Channel Islands, California. *Journal of biogeography* **10**:97–118.

Chapter Twelve: Southern California Mountains and Valleys

Borchert, M., A. Lopez, C. Bauer and T. Knowd. 2004. Field guide to coastal sage scrub and chaparral alliances of Los Padres National Forest. Technical Publication R5-TP-019. USDA Forest Service, Pacific Southwest Region, Goleta, CA.

Chambers, J. C., B. A. Roundy, R. R. Blank, S. E. Meyer, and A. Whittaker. 2007. What makes great basin sagebrush ecosystems invasible by *Bromus tectorum*? *Ecological Monographs* 77:117–145.

Conlisk, E., D. Lawson, A. D. Syphard, J. Franklin, L. Flint, A. Flint and H. M. Regan. 2012. The roles of dispersal, fecundity, and predation in the population persistence of an oak (*Quercus engelmannii*) under global change. *PloS ONE*: e36391.

de Gouvenain, R. C. and A. M. Ansary. 2006. Association between fire return interval and population dynamics in four California populations of Tecate cypress (*Cupressus forbesii*). *The Southwestern Naturalist* **51**:447–454.

Gordon, H. and T. C. White. 1994. Ecological guide to southern California chaparral plant series: Transverse and Peninsular Ranges: Angeles, Cleveland and San Bernardino National Forests. Technical Publication R5-ECOL-TP-005. USDA Forest Service, Pacific Southwest Region, San Diego, CA.

Lanner, R. M., and S. B. Vander Wall. 1980. Dispersal of limber pine seed by Clark's nutcracker. *Journal of Forestry* **78**:637–639.

Keeley, J. E. and F. D. Davis. 2007. Chaparral. Pages 339–366 *in* M. G. Barbour, T. Keeler-Wolf and A. A. Schoenherr, editors. *Terrestrial vegetation of California*, 3rd edition. University of California Press, Berkeley, CA.

Klein, A. and J. Evens 2006. Vegetation alliances of western Riverside County, California. Unpublished report to California Department of Fish and Game Habitat Conservation Division. California Native Plant Society, Sacramento, CA.

Minnich, R. A. 2007. Southern California conifer forests. Pages 502–538 *in* M. G. Barbour, T. Keeler-Wolf and A. A. Schoenherr, editors. *Terrestrial vegetation of California*, 3rd edition. University of California Press, Berkeley, CA.

Neel, M. C. 2000. The structure of diversity: implications for reserve design. Ph.D. thesis. University of California, Riverside, CA.

Stephenson, J. R. and G. M. Calcarone. 1999. Southern California mountains and foothills assessment: habitat and species conservation issues. General Technical Report GTR-PSW-172. USDA Forest Service, Pacific Southwest Research Station, Albany, CA.

Syphard, A. D., C. R. Volker, T. J. Hawbaker and S. I. Stewart. 2009. Conservation threats due to human-caused increases in fire frequency in Mediterranean-climate ecoystems. *Conservation Biology* **23**:758–769.

Chapter Thirteen: Mojave Desert

Brooks, M. L., and J. R. Matchett. 2003. Plant community patterns in unburned and burned blackbrush (*Coleogyne ramosissima* Torr.) shrublands in the Mojave Desert. *American Midland Naturalist* **63**:283–298.

Cole, K. L. 1986. The lower Colorado River valley: a Pleistocene desert. *Quaternary Research* **45**:392–400.

EPA. 2007. *Omernik Level III ecoregions of the continental United States (maps).* National Health and Environmental Effects Research Laboratory, Environmental Protection Agency (EPA), Washington, DC. Available at: http://www.epa.gov/wed/pages/ecoregions/level_iii_iv.htm. Accessed December, 2010.

MacMahon, J. A. 1979. North American deserts: their floral and faunal components. Pages 21–82 *in* D. W. Goodall and R. A. Perry, editors. *Arid-land ecosystems: structure, functioning, and management,* Volume 1: International Biological Program. Cambridge University Press, Cambridge, MA.

Keeler-Wolf, T. 2007. Mojave Desert scrub. Pages 609–656 *in* M. G. Barbour, T. Keeler-Wolf, and A. A. Schoenherr, editors. *Terrestrial vegetation of California*, 3rd edition. University of California Press, Berkeley, CA.

Norris, R. M. and R. W. Webb 1990. *Geology of California*. John Wiley, New York, NY.

Raven, R. H., and D. I. Axelrod. 1995. *Origin and relationships of the California flora.* California Native Plant Society, Sacramento, CA.

Ricketts, T. H., E. Dinerstein, D. M. Oldon, C. J. Loucks, W. Eichbaum, D. DellaSalla, K. Kavanagh, P. Hedao, P. Hurley, K. Carney, R. Abell, and S. Walters. 1999. *Terrestrial ecoregions of North America: a conservation assessment.* Island Press, Covelo, CA.

Thomas, K. A., J. Franklin, T. Keeler-Wolf, and P. Stine. 2004. *Mojave Desert ecosystem program central Mojave vegetation database.* Final report, U.S. Geological Survey, Western Ecological Research Center, Sacramento, CA, and Southwest Biological Science Center, Colorado Plateau Field Station, Flagstaff, AZ.

Turner, R. M., J. E. Bowers, and T. L. Burgess. 1995. *Sonoran desert plants: an ecological atlas.* University of Arizona Press, Tucson, AZ.

Webb, R. H., J. W. Steiger, and E. B. Newman. 1988. *The response of vegetation to disturbance in Death Valley National Monument, California.* Survey Bulletin 1793. USDI, U.S. Geological Survey, Washington, DC.

Webb, R. H., M. B. Murov, T. C. Esque, D. E. Boyer, L. A. DeFalco, D. F. Haines, D. Oldershaw, S. J. Scoles, K. A. Thomas, J. B. Blainey, and P. A. Medica. 2003. *Perennial vegetation data from permanent plots on the Nevada Test Site, Nye County, Nevada.* Open-File Report 03-336. USDI, U.S. Geological Survey, Tucson, AZ.

Chapter Fourteen: Sonoran–Colorado Desert

Fulton, R. 1984. Floral morphology and pollination in *Agave deserti* Englm. *Crossosoma* **10**:1–11.

Ingram, S. 2008. *Cacti, agaves, and yuccas of California and Nevada.* Cachuma Press, Los Olivos, CA.

Jefferson, G. T. and L. Lindsay. 2006. *Fossil treasures of the Anza Borrego Desert.* Sunbelt Publications, San Diego, CA.

Killinbeck, K. T. 1990. Leaf production can be decoupled from root activity in the desert shrub ocotillo (*Fouquieria splendens* [Engelm.]). *American Midland Naturalist* **124**:124–129.

Marks, J. B. 1950. Vegetation and soil relations in the lower Colorado Desert. *Ecology* **31**:176–193.

Mathias, M. E. 1978. The California Desert. *Fremontia* **6**:3–6.

Nobel, P.S., editor. 2002. *Cacti: biology and uses.* University of California Press. Berkeley, CA.

Oglesby, L. C. 2005. The Salton Sea. *Memoirs of the Southern California Academy of Sciences,* Volume 10. Southern California Academy of Sciences, Los Angeles, CA.

Rebman, J. T. and N. C. Roberts. 2012. *Baja California plant field guide,* 3rd edition. San Diego Natural History Museum and Sunbelt Publications, San Diego, CA.

Schoenherr, A. A. 1992. *A natural history of California.* University of California Press, Berkeley, CA.

Schoenherr, A. A. and J. Burk. 2007. Colorado Desert vegetation. Pages 657–682 *in* M. G. Barbour, T. Keeler-Wolf, and A. A. Schoenherr, editors. *Terrestrial vegetation of California,* 3rd edition. University of California Press, Berkeley, CA.

Shreve, F. 1925. Ecological aspects of the deserts of California. *Ecology* **6**:93–103.

Shreve, F., and I. L. Wiggins. 1964. *Vegetation and flora of the Sonoran Desert,* Volume 1. Stanford University Press, Stanford, CA.

Solbrig, O. T. and G. H Orians. 1977. The adaptive characteristics of desert plants. *American Scientist* **65**:412–421.

Tevis, L., Jr. 1958. Germination and growth of ephemerals induced by sprinkling a sandy desert. *Ecology* **39**:681–688.

U. S. Bureau of Land Management. 2002. Proposed Northern and Eastern Colorado Desert coordinated management plan. Prepared by Department of the Interior Bureau of Land Management, California District Office and California Department of Fish and Game Inland, Desert and Eastern Sierra Region.

Zabriskie, J. G. 1979. *Plants of Deep Canyon.* Philip L. Boyd Deep Canyon Desert Research Center, University of California, Riverside, CA.

Index

USDA-NRCS. 2015. The PLANTS Database. National Plant Data Team, Greensboro, NC. Available at: http://plants.usda.gov. Accessed September 2015.

University and Jepson Herbaria (UC/JEPS). 2015. University of California, Berkeley, CA. Available at: http://ucjeps.berkeley.edu/interchange. Accessed September 2015.

Index

TO PLANTS* IN THIS BOOK

* Primary Source: USDA-NRCS (2015)
Illustrations indicated by bold italics

A

B

C

D

E

F

N

O

P

T

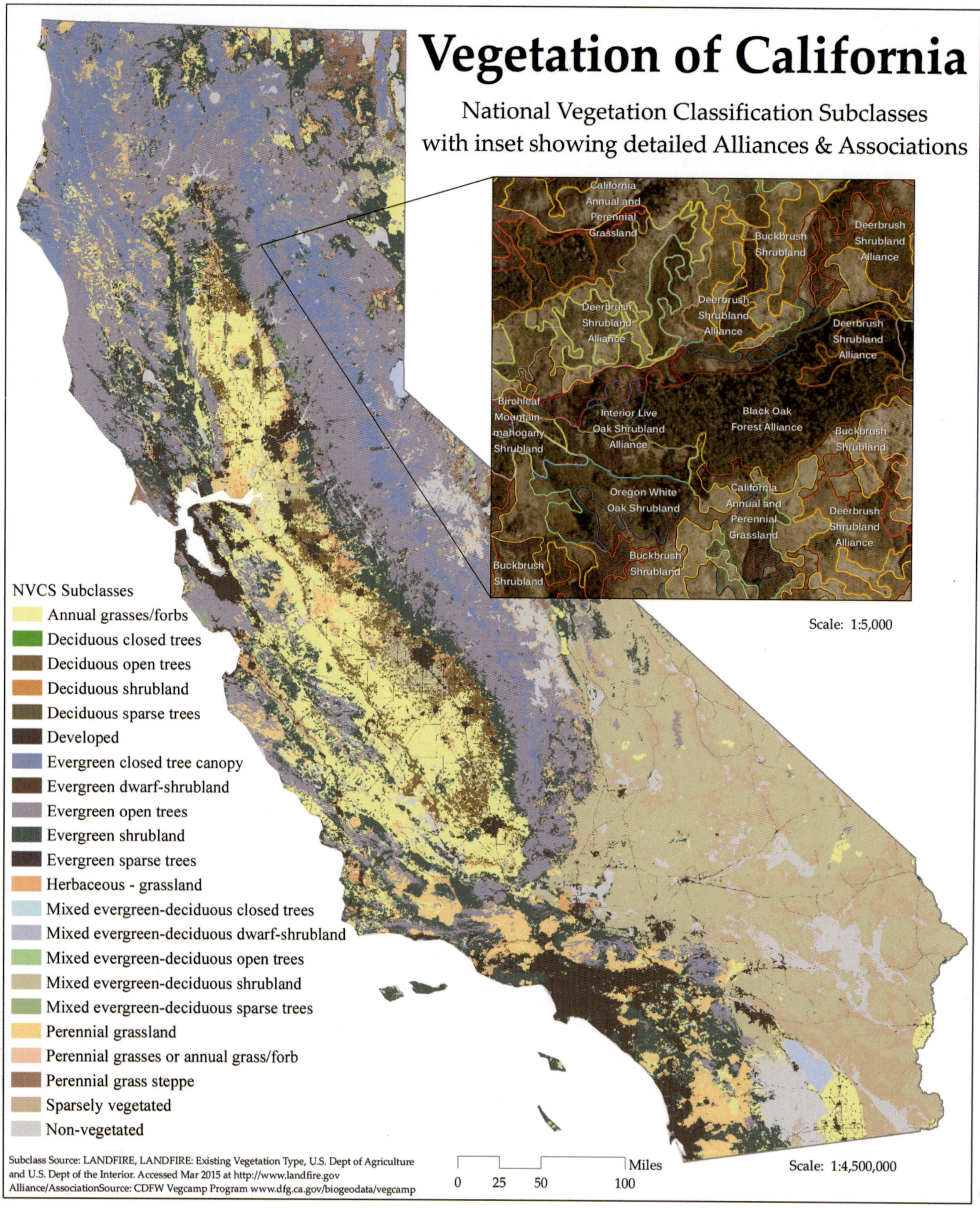

Charlie Frye & Charles Convis/esri

Vegetation, a natural component of ecoregions, can be organized and depicted in maps from coarse- to fine-resolution. This statewide map displays 21 coarse-resolution categories useful for general discussion and for comparing vegetation around the world. The inset map portrays a more detailed level of individual map polygons, which match the patterns of vegetation seen in the aerial imagery backdrop. Detailed maps have been produced for over a third of the state, using definitions of field data that represent repeating floristic patterns. Such fine-resolution maps are ideal for conservation groups and land managers to use in biotic inventories for habitat assessment, fire threat assessment, biomass depictions of carbon storage, and landscape connectivity analyses for wildlife, to name just a few possible uses.

From almost anywhere in California, you will find a diverse and unique array of vegetation, underscoring the effects of local environments on the varied combinations of thousands of plant species. We hope this book reinvigorates your passion to study, enjoy, and protect our state's incredible richness of natural landscapes.

California Native Plant Society